ENGINE MANAGEMENT
Advanced Tuning

GREG BANISH

CarTech®

CarTech®

CarTech,® Inc.
6118 Main Street,
North Branch, MN 55056
Phone: 651-277-1200 or 800-551-4754
Fax: 651-277-1203
www.cartechbooks.com

Edit by Josh Brown

ISBN 978-1-932494-42-6
Item No. SA135

Written, edited, and designed in the U.S.A.
Printed in China
15

Title Page:
The inertial chassis dyno is perhaps the most common dyno used today. Low cost and easy operation make them an attractive tool to the aftermarket. (Nate Tovey)

Back Cover Photos

Top Left:
The modern hot wire element MAF sensor is an extremely accurate method of measuring incoming air mass. Notice the honeycomb flow element and wire screen that help reduce turbulence entering the meter.

Top Right:
The addition of full electronic controls gives the calibrator much greater flexibility in the tuning process. What used to take hours of dirty work can now be accomplished in minutes with just a keyboard and the right software.

Middle Left:
Exhaust oxygen sensors shown with (foreground) and without internal heaters. The addition of the heating elements allows for accurate measurement a shorter time after engine startup.

Middle Right:
The cold air kit installed in front of the factory MAF alters the way air flows into the sensor. The new MAF output versus actual airflow must be entered into the EEC for proper fuel, load, and spark calculations.

Bottom Left:
Most PCMs use speed and load for inputs to the base ignition map. The actual delivered timing may vary from this if other adders for temperature, lambda, or torque control are active.

Bottom Right:
Modern PCMs pack as much processing power as the desktop computers of several years ago. In the time it took you to read this caption, this processor can perform over 15 million calculations.

DISTRIBUTION BY:

Europe
PGUK
63 Hatton Garden
London EC1N 8LE, England
Phone: 020 7061 1980 • Fax: 020 7242 3725
www.pguk.co.uk

Australia
Renniks Publications Ltd.
3/37-39 Green Street
Banksmeadow, NSW 2109, Australia
Phone: 2 9695 7055 • Fax: 2 9695 7355
www.renniks.com

Canada
Login Canada
300 Saulteaux Crescent
Winnipeg, MB, R3J 3T2 Canada
Phone: 800 665 1148 • Fax: 800 665 0103
www.lb.ca

ACKNOWLEDGMENTS

This book would not have been possible without a lot of help through the years. Many of the relationships I've made both personally and professionally have been fuel to keep me going over the years and opened the doors that made this all possible.

I would like to thank the people at SCT, Mustang Dynamometer, Dynojet Research, HPTuners, Engine Controls and Monitoring, Mr. Gasket, F.A.S.T., Vortech Engineering, MRT Racing, SiemensVDO, Magnuson Products, Diablosport, MSD, Autronic, Trick Flow Specialties, Ford, and Mopar for their recent help and opportunities to challenge myself over the years.

Nate Tovey, my photographer for this book, put in plenty of long hours alongside me when it counted. Thanks buddy, we made it happen.

Andrea, my wife, put in just as many long hours as I ever did. You stood by me while I chased my dream and earned my experience the hard way: one long night at the shop after another. Not only have you gracefully put up with my obsessions, but you did it with a smile. You've always been there for me and helped me keep my focus. What more could a guy ask for?

Dave and Jay, if only you could join me now for the good times. Godspeed and God bless.

–Greg Banish

ABOUT THE AUTHOR

Greg Banish is a mechanical engineer and motorsports enthusiast who works as an OEM calibrator by day and professional engine tuner in his spare time. Greg earned a bachelor's degree in Mechanical Engineering at GMI Engineering & Management Institute (Kettering University) with a specialty in automotive applications, engine design and power systems. He wrote his thesis on vehicle instrumentation and wheel torque measurement.

Putting this education to practical use, he founded his own performance shop outside of Detroit and has served local enthusiasts, shops, automotive companies and OEMs. A background in engineering drove him to approach engine tuning logically by applying an OEM-style approach. As one of the earliest proponents for using a load bearing chassis dyno for general enthusiast

Author Greg Banish with his dog, Turbo.

tuning work, he worked to realign the public view of how performance tuning should be done. With over a thousand unique aftermarket calibrations performed on the dyno and "track," he has had the opportunity to work with a wide variety of engines and control systems.

Currently working as a calibration engineer, Greg works closely with engineers from U.S. and European OEMs including environmental testing in some of the most extreme conditions. His assignments have covered engine control technologies ranging from torque-based controls to thermal modeling to gasoline direct injection. He is also a member of the Society of Automotive Engineers and SEMA.

To this day, he continues to serve the performance industry providing EFI calibration and consulting services to shops as well as teaching seminars on engine calibration that are available to both professional tuners and individual enthusiasts alike. He enjoys road racing his Ford Mustang and instructing at open track events.

INTRODUCTION

Before this book even begins, I wish to make it perfectly clear that this is not an engine design or combustion theory text. The goal here is for the educated enthusiast, skilled technician, and automotive engineer alike to all be able to come away with something. To this end, we explore the basics of engine operation to reinforce what is really going on under the hood. From there, we move on to the "ins and outs" of modern electronic fuel injection systems and ultimately some specifics of calibration methods and horsepower

The 1968 Plymouth GTX 440 made 375 gross hp and was considered state of the art for its time. In reality, this engine has a specific output of 0.85 hp per cubic inch and would fail modern emissions tests miserably. (Nate Tovey)

The 2006 Ford Mustang makes 300 hp out of a 281 cubic engine for a specific output of 1.07 hp/ci, with complete compliance to today's stringent emissions standards. (Nate Tovey)

The engine in this Mercedes 220SE is equipped with mechanical fuel injection and makes about 120 hp from 2.2 liters. It has limited capability to adjust for changing weather conditions. (Nate Tovey)

production. The focus of this book is gasoline engines; however, many of the concepts can carry over to other applications. While much of the material may seem like a review to many, it is important to keep in mind the fundamentals of engine operation while attempting to change calibrations. A solid understanding of what is happening inside the manifold and combustion chamber gives the calibrator an edge in tuning.

Let's face it, today's performance enthusiast doesn't want to compromise. We want tons of power, reliability, drivability, and worry-free operation. Gone are the days of living with the compromises between the horsepower seekers and the emissions regulators. We now live in a time where one can walk into a new car dealership and simply buy an honest 400-horsepower car that idles quietly, drives smooth as silk, and is backed by a full factory warranty. Considering that in the heyday of the muscle car wars 300 gross horsepower was astounding and it still came with a rough idle and terrible gas mileage, today's performance car market is as good as it has ever been.

So how did we get here? First and foremost, the automakers have learned a thing or two about engine design in the last three decades. Serious advances in the areas of cylinder head, intake, and camshaft design have allowed engines to make far more power out of much smaller packages and displacements. What the OEM engineers call specific output, or power per cubic inch, has gone way up directly as a result of the increase in flow potential of modern component designs. Compare today's injection molded long runner intake with the cast-iron four-barrel paperweight of the '60s and it's easy to see the difference. Other than the obvious weight advantage, the port walls are smoother and sizes are tuned to take advantage of standing waves to increase port energy at the same time the valves are opening. Friction has become another area where modern engines have evolved tremendously. Where there were once solid tappets dragging across an oiled cam under severe spring pressure, there are now hydraulically damped rollers—or even the complete lack of pushrods—shortening the path between the cam lobe and valve. Looking at a modern cylinder head also reveals carefully designed port geometry, combustion chambers designed to do more than simply seal ports, and often a camshaft or two. The head ports themselves have evolved to increase velocity, yielding more total flow through smaller valves and better mixing of the air and fuel in the combustion chamber.

With that said, efficiently designed air pumps don't run as a working engine without a little help. Current production vehicles run on electronic fuel injection for a whole list of reasons, not the least of which are emissions and drivability. Fortunately, emissions and power production are not completely at odds with one another as the environmentalists would have us believe. The underlying connection is efficiency. Taking advantage of every drop of fuel in the engine leaves less left over to pollute our precious atmosphere and ensures that we're not missing out on our chance to use the energy in that fuel to push as hard as possible on the piston to move us

This modern engine, although still carbureted, produces large amounts of power. But even at over 1,600 hp, it must still be adjusted to accommodate changes in current weather conditions for best performance. (Nate Tovey)

This Super Street Outlaw engine uses two sets of injectors, each on its own rail, to supply enough fuel to make almost 2,000 hp. With an EFI control system, changes in weather are automatically compensated for by the PCM to keep the engine running at its peak. (Nate Tovey)

From left to right: GM GENIII, Siemens SIM90, and Ford EEC-V. All are different packages for the same general set of control functions. (Nate Tovey)

Modern PCMs pack as much processing power as the desktop computers of several years ago. In the time it took you to read this caption, this processor can perform over 16 million calculations. (Nate Tovey)

down the road. The balance is to make sure that we only inject enough fuel to make the power necessary to do whatever it is we're asking the engine to do at the moment. Whether it's idling at a stoplight, cruising the interstate, or racing down the quarter mile, there is always an ideal recipe of air and fuel to pour into the engine to keep things working as close to peak efficiency as possible. The closer we can keep the engine to this ideal mix at all times, the better the engine will perform.

Sure, that old small-block Chevy in the garage has run great for years with a single Holley 650 on top and the timing locked at 34 degrees. It makes all the power you need to go out on Friday nights and cruise the memory lane of your choice to your satisfaction on any summer night. So how does it start in March? Why do you need to re-jet the carburetor to run good every October? Why did that new Mustang at the last stoplight walk all over you? How can he do that to your hot rod and still be 100 percent emissions legal?

The answers come in the form of precision. The ability to instantly adjust to changes in conditions is what gives an electronically fuel-injected car the decided advantage here. Carburetors can be tuned perfectly under any single condition. This is why you see drag racers constantly checking their weather stations in the pits between rounds and making adjustments to the jets. With a change in density comes a change in what would be considered an ideal mixture at any given moment. What was spot-on last night can now be five percent rich this afternoon. The beauty of EFI is that changes to the engine's operating parameters, necessary to keep up with a changing atmosphere, happen without even lifting the hood. Further, the changes that EFI can make are in smaller increments than what can be done with a jet change on a carburetor. It's not unusual to see adjustments in half a percent of fuel delivery or even a quarter of a degree of spark lead in a modern fuel injection system. The whole trick is to know where to have the controller make these adjustments.

The speed at which the electronic processors inside modern Powertrain Control Modules (PCMs) operate is extremely fast, even when compared to the operating speed of an engine at redline. Processor speeds are measured in Megahertz, or millions of cycles per second. Even the most advanced Formula 1 engines today only operate at a maximum of approximately 23,000 rpm. It becomes easy to see how the PCM has plenty of time to think about what outputs to deliver. The PCM can take a snapshot of the engine's performance at any instant, analyze every parameter, and make multiple calculations before the crankshaft even rotates a couple of degrees. This process can be repeated multiple times between combustion events at high speed.

In the early years of electronic fuel injection, control of system and engine operation was reserved for OEM engineers. The added complexity did a good job of scaring many performance enthusiasts away. Little by little the aftermarket began to step up to the plate and embrace the technology. Eventually enthusiasts were rewarded with replacement EPROM (Erasable Programmable Read Only Memory) chips modified to provide the subtle changes to air/fuel mixture and timing needed to squeeze 10 or more horsepower out of production vehicles. After that came the aftermarket's trump card, the stand-alone fuel injection system. No longer were performance enthusiasts tied to factory fuel and timing maps. New, more exotic combinations of parts could be employed in the name of horsepower.

With the advent of the Internet, we have seen hundreds of enthusiast sites spring up dedicated to just about any performance make, engine, and vehicle imaginable. The question that seems to get echoed across all forums seems to be: "How do I change the PCM to get more out of my car?" Not unlike politics, everyone seems to have their own opinion of what's best, and nobody seems shy to argue that they are right. While some forums seem to be far more technically oriented and mindful of avoiding the proliferation of misinformation, there seems to be no shortage of armchair quarterbacks in the Internet tuning arena. The best advice here seems to be to take everyone with a grain of salt. It is rare to see true experts online teaching the general public the secrets to their success for free. More often it's the grassroots enthusiasts sharing what tip or trick most recently worked for them.

The word of caution here is to remember that with the complexity of today's EFI systems it's easy to get what seems like the right result for the wrong reason. Many output functions are controlled by layer upon layer of tables, scalars, and functions that work in concert to deliver the seamless operation of the engine. Changing the wrong function or table may indeed correct the problem at hand while unknowingly creating another. The benefit of the EFI system is that the layers of control, if employed correctly, can independently prevent individual performance issues while maintaining a complete package that drives smoothly. Don't feel bad if this seems intimidating at first; it takes time to fully understand the layers of control in an EFI system. Many systems use slightly different strategies or naming to describe and control the same physical systems. With a little bit of patience and attention to detail, it is possible to tame even the most complicated system.

This book covers a logical approach to EFI calibration. The methods discussed here closely follow those used by OEM calibration engineers on current production vehicles. We cover a solid foundation of proper setup, airflow modeling, and fine-tuning of drivability parameters. This approach works regardless of the EFI system being used. Whether you are reflashing the PCM in a 2005 OBD-II equipped BMW, burning a new chip for a 1992 Mustang, or tuning a dedicated racecar using an Accel DFI GenVII standalone processor, the tuning procedure is fundamentally the same. I encourage the calibrator to constantly think about what is going on inside the combustion chamber when making adjustments. Understanding how much air is making its way into the cylinder and how fast the mixture is burning after ignition goes a long way toward making the right decision when changing engine maps during the calibration procedure. At the end of the day, almost all EFI systems operate on the same set of principles. The processor does not know what kind of car it is running, only what inputs and outputs it has. The same laws of physics and thermodynamics apply to all engines equally. A proficient calibrator can tune any engine as long as he has access to the necessary hardware and software.

Known by many names, engine controllers are still engine controllers. It seems as though almost every manufacturer has its own name for their electronic controller. Examples include Powertrain Control Module (PCM), Engine Control Module (ECM), Engine Control Unit (ECU), and Electronic Engine Controller (EEC). For the sake of simplicity, we use PCM through this text as a generality

THE BASICS

Before we delve into the finer points of EFI table adjustments, let's take a step or two back and understand exactly what is happening under the hood. It has been said before and it is reinforced here again: An engine is little more than an air pump. It just so happens to be that with the right mix of ingredients, we get to harvest a little hidden energy on the way through. Both the air we breathe and the gasoline in the tank are made up of complex combinations of chemicals with all kinds of hidden potential. However, this reaction can only take place between certain amounts of each chemical. Just like a baker knows that too much flour turns cookies into biscuits, too much fuel gives us a less-than-ideal reaction.

Ideally, to have no extra molecules of either oxygen or fuel left over at the end of the reaction, we must start with the right ratio of components. Chemists call this correct ratio a stoichiometric mix. For gasoline and air, it is 14.68 pounds of air for every pound of gasoline. Notice that we say pounds of air and not cubic feet. On a molecular level, each string of octane and each oxygen molecule have a specific mass. To get the right ratio of strings of octane to oxygen molecules, we must calculate based on mass. Changes in barometric pressure, manifold pressure, and temperature have a significant impact on the density of air and fuel, so we must remember that one cubic foot of air does not always contain the same number of oxygen molecules. Once again, this is where electronic fuel injection shines with its ability to compensate for such changes almost instantaneously.

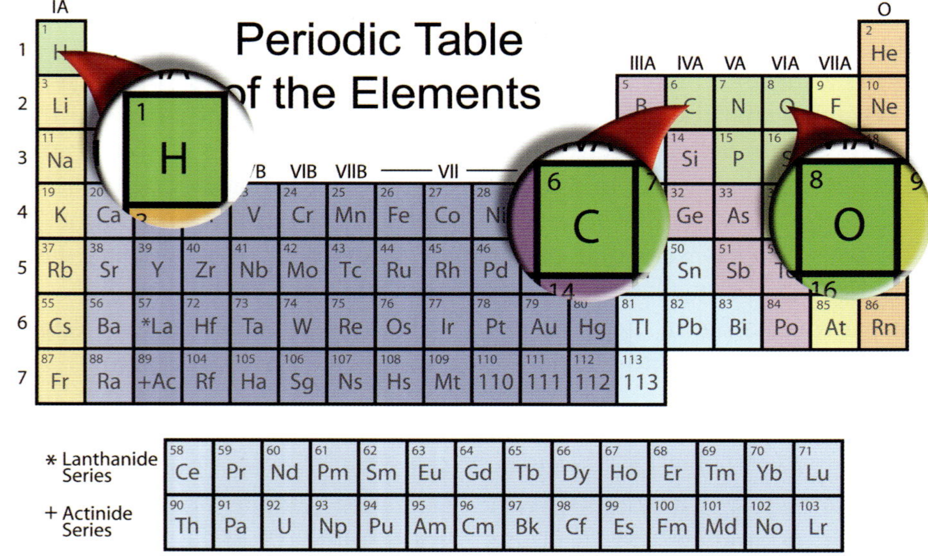

From the periodic table of elements, we can see the molar weight of oxygen, hydrogen, and carbon. When these combine to form isooctane or O2 molecules, the mass of each can be calculated from the molar weight of its components. (Nate Tovey)

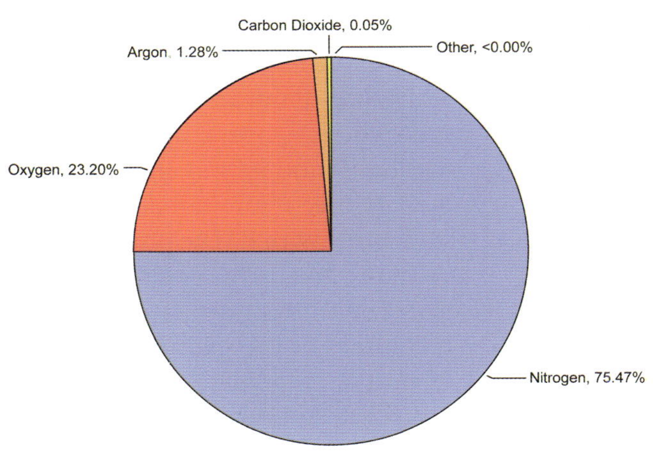

The air you are breathing right now is not pure oxygen. The primary component is actually nitrogen, with only about 23% oxygen. This is why the engine must consume more total air mass to obtain the oxygen required to react with the gasoline. (Nate Tovey)

4 Cycles of an Engine

"Suck, Crush, Bang, Blow," or at least that's how I remembered it in college. Maybe "Intake, Compression, Power, Exhaust" is less offensive, but I doubt it leaves as much of an impression on the mind.

The process begins with a relatively empty space inside the cylinder of the engine and an open intake valve. As the piston moves down, pressure inside the cylinder drops below that of the intake tract and atmosphere. This pressure difference is what pushes the air and fuel into the chamber. Since we know that each ounce of fuel only carries so much energy and must be mixed with an appropriate amount of air to burn, the more total mix that can find its way in to the cylinder each time the valve opens, the more potential we have to make power.

Obviously there are ways to increase or decrease the amount of charge filling the cylinder each time. The most obvious method of charge fill control is the throttle blade. By closing off a portion of the inlet tract with a blade, the amount of air available for the next intake stroke is reduced. Mixing a smaller amount of intake air with a smaller amount of fuel yields a smaller power potential, but also resists the tendency of the engine to pick up speed. Conversely, with a throttle blade completely open, the amount of air entering the cylinder can be increased by taking advantage of standing waves in tuned length runners, leaving the valve open longer with changed cam timing events, or changing the pressure differential severely with supercharging.

Actual charge fill is an indicator of how much work the engine is doing at the moment. The more charge filling the cylinders, the harder the engine is working. This work is expressed in terms of "load" or "Volumetric Efficiency (VE)." Load and volumetric efficiency are just two methods used to describe the actual mass of airflow through an engine compared to the theoretical mass flow based on its displacement and speed. The theoretical amount of charge fill is the mass of air that would occupy the same volume as the engine displaces. This mass is found by multiplying volume and normal atmospheric density.

$$M_{air,engine} = (V_{engine}) \times (\rho_{air})$$

The theoretical filling is calculated at standard temperature and pressure (STP or STD) where density (ρ_{air}) is equal to 0.00004671 lb_m/in^3. To find the flow rate, we normalize for the number of complete displacements over time. A four-stroke engine has two revolutions per cycle of displacement, so the displacement rate is one-half of actual engine speed. The theoretical airflow rate of an engine is:

$$\dot{M} = (V_{engine}) \times (\rho_{air}) \times (RPM/2)$$

The theoretical 100% filling of a four-stroke engine with a 300 cubic inch displacement operating at 1000 rpm is calculated by:

$$\frac{(300\ in^3) \times (0.00004671\ lb_m/in^3) \times (1000\ rpm)}{2}$$
$$= 7.01\ lb/min\ air\ mass\ flow\ through\ engine$$

This engine would theoretically move 7.01 pounds per minute of air through itself at standard ambient conditions and 1,000 rpm. By closing the throttle blade to reduce airflow and horsepower, the engine is only allowed to move 1.40 pounds per minute. The ratio of actual instantaneous mass flow to theoretical pumping gives us volumetric efficiency or load.

$$\frac{(1.40\ lbs/min)}{(7.01\ lbs/min)} = 0.20\ or\ 20\%$$

This means that the reduced throttle position yields a volumetric efficiency of 20%. Typical engine loads can vary from 10 to 18% at idle. Most steady cruising at street and

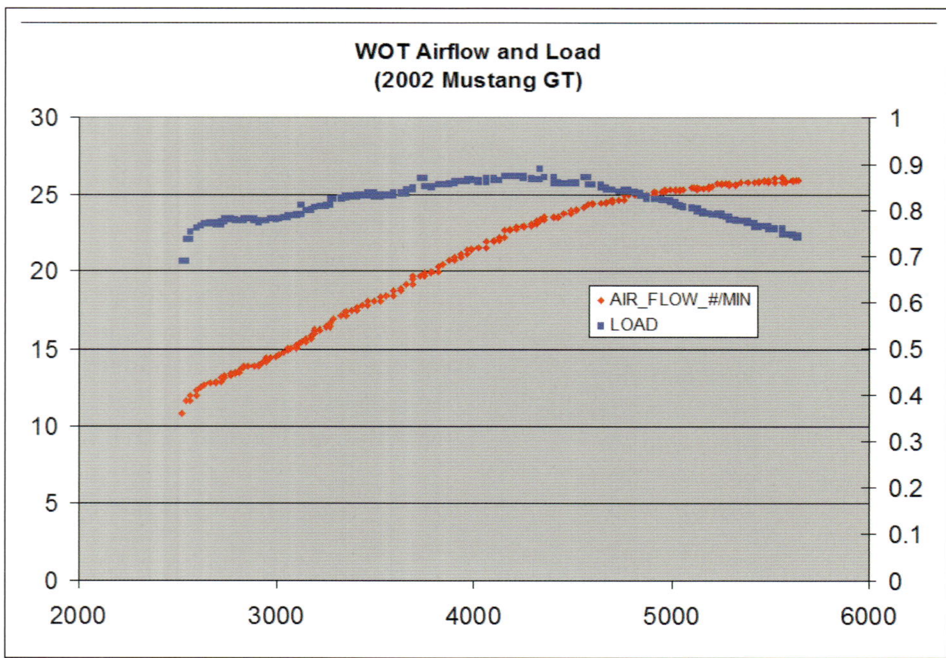

Datalogging of the engine sensor data during a wide open throttle run reveals the actual air consumption. This air consumption is used by the PCM to calculate engine load. Notice how this naturally aspirated engine develops approximately 86% engine load at WOT.

highway speeds happens at approximately 20 to 30% load. Light acceleration is usually between 30% and 60% engine load. Wide-open throttle operation for naturally aspirated engines results in a load anywhere from 60% to 105%, depending on how efficient the intake, camshaft, and cylinder designs are. Supercharged engines routinely see loads in excess of 100% once boost is present in the manifold. For reference, most 5.0L Ford Mustang engines show approximately 80% load at peak power. With a Vortech centrifugal supercharger making 8 psi of manifold boost, the same engine can show approximately 140% load at peak power.

EFI systems excel by being able to accurately calculate engine load at any time. Load has a large impact on what the engine wants for operating parameters to perform optimally. As load increases, spark advance typically has to be reduced to prevent knock. At high loads, fuel enrichment may also be necessary to control exhaust gas and component temperatures.

Any way you slice it, just pumping air from the outside world into the cylinder takes some amount of energy. The engine needs to come up with some power to drive this process somewhere. Once a mix of air and fuel has found its way into the cylinder, we need to find a way to do something with it. The piston moving up in the bore with both valves closed does two things very effectively. First, it compresses the mixture, giving it a new set of properties. A denser air/fuel mix tends to put all of the molecules in much closer proximity, making for a faster reaction once things start to happen. Also, Boyle's Gas Law tells us that as the volume decreases, pressure and temperature increase. $PV=nRT$, remember? A hotter, denser mix makes for a faster reaction. Just like the intake stroke, compressing this mixture takes energy to drive the piston up against a mixture that generally doesn't want to get any smaller on its own. This energy needs to come from somewhere too.

So where does the power come from? Hydrocarbons in the fuel have a tendency to give off lots of heat

An early form of fuel injection is central port injection. Seen here, the injectors are mounted in the middle of the intake manifold on what essentially looks and acts like an electronic carburetor. (Nate Tovey)

and pressure when mixed with the right amount of oxygen from the air in the presence of heat. Engineers and chemists call this an exothermal chemical reaction resulting from the mixture of octane and oxygen with a heat catalyst (pyrolysis); we call it the combustion. This combustion releases heat that dramatically increased the pressure in the cylinder. The real trick for the calibrator is to create conditions in which as much of the air/fuel mix that has already been introduced into the cylinder reacts during the time when the piston is traveling downward. Taking advantage of 100% of the available chemical energy in the charge mix during this time means less wasted energy to be dumped out the exhaust or simply not react at all, yielding better emissions and power output. Once we have had this reaction that releases heat and pressure, the result is a piston that is pushed down the bore giving rotational energy to the crankshaft. If the energy pushing down on the piston in the power stroke equals the energy needed to pump the air through the engine, compress the mix, overcome friction, and rotate the accessories on the engine, speed stays constant. If the energy resulting from the power stroke exceeds the other losses, the engine accelerates. Likewise, less energy pushing on the piston than current losses yields a deceleration.

At the end of the power stroke, we are left with a cylinder full of mostly spent gases. As the exhaust valve opens, the piston drives upward providing enough pressure differential to push things out. Obviously, the greater this pressure differential is between the cylinder and exhaust system, the more flow.

Much like the intake manifold, proper tuning here to take advantage of standing waves helps scavenge the cylinder more efficiently, which leaves room for more air and fuel on the following intake stroke. As a side benefit of the relatively high pressures driving exhaust gases out of the cylinder, it often becomes rather easy to increase total system flow by reducing restriction in the exhaust system as a whole. The major trade-off here is noise. This is also where a turbocharger shines with its ability to harness the high pressure, high temperature exhaust energy for more help on the intake side. We will cover more about this later.

Air and Fuel

Let's take a look at what we're really moving through the engine for a moment. "Air" is actually a mix of primarily nitrogen and oxygen. The ultimate goal of the EFI system is to accurately determine the correct ratio between oxygen being ingested by the engine and fuel being delivered through the injectors. In the science of calibration, we must make the assumption that our external environment is a relatively constant mix of approximately 23% oxygen by mass (20% by volume). Even as ambient conditions change the density of the air, all we can really do is model total air mass entering the cylinders.

Oxygen is the real player in the combustion event, since it is the molecule that eventually mixes with the hydrocarbon to yield carbon dioxide and water. Yes, water. Keep in mind that under normal operating conditions, exhaust gas temperatures are well in excess of water's boiling point of 212 degrees F, yet water is still present. It takes rela-

tively cold ambient conditions to actually see steam in the exhaust. On cold starts, the exhaust system itself on most vehicles is cool enough to allow some steam to condense to a liquid. An interesting side note is that repeated short trips where the exhaust is cooled below this condensation point yield a fair amount of liquid water, often sitting inside the muffler. This can sometimes lead to corrosion of unprotected components.

The primary component of our atmosphere, nitrogen, is mostly just along for the ride during the whole engine cycle as far as power is concerned. However, it is important to note that emissions are directly impacted if combustion temperatures get high enough to break down the nitrogen atoms, yielding oxides of nitrogen in the exhaust. To combat this, Exhaust Gas Recirculation (EGR) is often employed. Recirculating spent gases back into the intake charges dilutes the energy potential of the incoming charge, reducing its burning temperature potential. Additionally, unburned oxygen and fuel molecules have a second chance at properly mixing before being wasted out the tailpipe. EGR is often controlled by a valve directly between the exhaust system and intake tract. Another source of EGR is camshaft overlap. Whenever both valves are open simultaneously, there is an opportunity for intake and exhaust gases to mix in both the cylinders and intake ports. This effect is typically more pronounced at low speeds where ram tuning has not yet become dominant, but it should be kept in mind for later tuning concerns.

With many different hydrocarbon fuels to choose from, what's the difference? A major consideration is the amount of chemical energy

---

Fuel Type	Lower Heating Value	Octane	Density	Stoichiometric Ratio
	kJ/kg	(R+M)/2	kg/l	kg/kg
Gasoline				
Regular	42.7	87	0.740	14.8
Premium	43.5	93	0.755	14.7
Iso-Octane	44.8	100	0.690	15.2
Diesel	42.5	25	0.835	14.5
Methanol	19.7	104.5	0.790	6.47
Ethanol	26.8	104.2	0.790	9
E85	29.19	101.6	0.783	9.87
Propane	46.3	104.5	0.510	15.67
Hydrogen	120	130	0.090	34.3
Methane	50	120	0.720	17.2
Kerosene	43	15	0.800	14.5
Benzene	40.2	115	0.880	13.3
Toluene	40.6	114.5	0.870	13.4
LPG	46.1	110	0.540	15.5

Figure 2-1 *Heating values for several common fuels. Notice that the energy contained in these fuels is not always proportional to the amount required for stoichiometry. (Nate Tovey)*

stored in the fuel molecules. This energy value of the fuel is known as its "heating value." The higher this number, the more total energy is hidden in the fuel molecules. Good fuels by definition contain plenty of this chemical energy potential and can be consistently mixed and ignited inside of an engine. The heating value is expressed in terms of energy per unit mass (kJ/kg, BTU/lb), making it easy to see how burning more fuel mass releases more energy (see **Figure 2-1**, properties of fuels). Gasoline is a good automotive fuel choice due to its relatively high heating value, acceptable boiling point, ease of storage, and relative ease of production.

Much like the air in the atmosphere, modern fuels are cocktails of varying ingredients. Gasoline itself is not a pure fuel, but the blend most closely resembles the properties of isooctane. (**Figure 2-2**) As the name suggests, octane is a chain of eight carbon atoms each carrying a pair of hydrogen atoms and one more hydrogen atom to cap each end of the chain (C_8H_{18}). This makes for a relatively long molecule with plenty of exposed single bonds. It is the breaking of these bonds and reforming of new ones during combustion that generates the heat energy driving the power stroke. Isooctane has three of the end clusters of this chain moved toward the center, making it a bit more stable molecule, requiring more energy to break.

Figure 2-2 *Isooctane is the closest single molecule that best represents the properties of gasoline. (Nate Tovey)*

Fuels with a higher octane number have a stronger concentration of these more robust molecules versus the more easily broken longer chains. Thus, they require more energy to break these bonds, and have more resistance to knock. An interesting side note is that the energy required to break these bonds comes in the form of heat during combustion. Heat transfer from one molecule to the next requires a small amount of time. Even though this is a very small amount of time, more total transfer means more time for it to happen. Adding more heat to the molecules in order to separate their ions means more time is required. This means that the burn rate is slightly slower for "high octane" fuels. This becomes important later when we choose spark advance timing.

We have already covered the basic recipe for combustion, but much like any family recipe there can be variations. Stoichiometric combustion of gasoline occurs at a ratio of 14.68:1, but combustion can occur anywhere between 7.5:1 and 26:1. This leaves a broad range of possible operating conditions. The catch is that this chemical reaction releases differing amounts of heat depending on the mixture. Generally speaking, rich mixtures (excess fuel, $\lambda < 0.9$) burn cooler with more hydrocarbon emissions, and slightly lean mixtures (excess air, $\lambda \approx 1.05$) burn hotter with oxides of nitrogen forming as a direct result. For emissions, we target a stoichiometric mixture to leave as few byproducts either way. (**Figure 2-3**)

Engineers often refer to the air/fuel ratio in terms of "Lambda" (λ). Lambda is defined as an excess air ratio. Lambda is equal to 1.00 exactly at stoichiometric mixture and increases as the air/fuel ratio gets

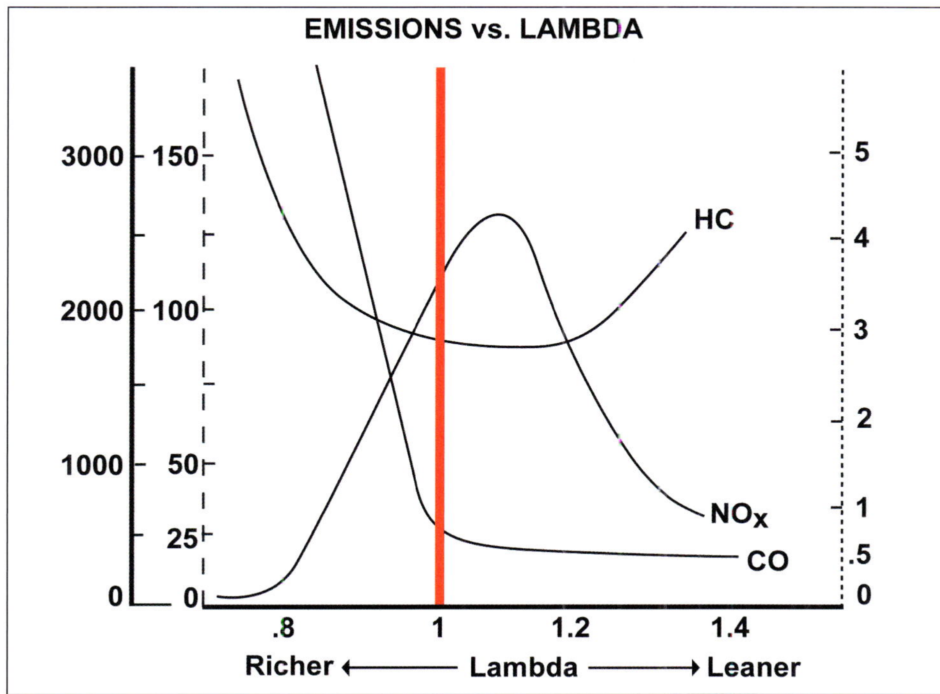

Figure 2-3 *Lambda influence on component emissions. Running either rich or lean of stoichiometry will drive emissions up. Burn temperature control is the preferred method of controlling NOx emissions. (Nate Tovey)*

leaner. Most OEM code is written in terms of lambda for fuel control. This gives the calibrator a quick percentage reference of their deviation from stoichiometric mixture. Modern engines are run at stoichiometry ($\lambda = 1$, 14.68:1 A/F) under most conditions to balance emissions and fuel economy, so calibrators spend a lot of time targeting this ratio. Lambda is a convenient number to work with since its value represents the correction necessary to the airflow calculation in order to return to stoichiometry.

For best power, it has been found that somewhere between 13.2 and 13.4:1 ($\lambda \approx 0.95$) is ideal. (**Figure 2-4**) The slight excess of fuel in this mix means that enough extra fuel has been injected to ensure that as many of the oxygen molecules as possible react with available fuel to generate power. Since the total power limit of most engines is total

mass airflow capacity, this means the engine uses every bit of air it digests in the name of power. Often we find that this ratio is simply not possible

> ### Equivalence Ratio
>
> Another term often used to describe air/fuel ratio is "Equivalence Ratio." Equivalence ratio is the inverse of lambda, with higher numbers representing richer conditions. Think of this as percent enrichment. Equivalence ratio is often used to describe power enrichment or catalyst cooling conditions. Simply put, for gasoline engines:
>
> λ = (Actual air/fuel ratio)/ (14.68)
>
> EQ Ratio $\varphi = 1/ \lambda = (14.68) /$ (Actual air/fuel ratio)

to run with the onset of knock in forced induction applications. To offset this, a richer mixture is run for safety, but the phrase is coined: "leaner is meaner." Most naturally aspirated engines with conservative ignition timing exhibit a bell curve of power centered around this ratio. Smart calibrators stay on the richer side for safety in a normal driver or racecar. In some rare instances such as endurance racing, the leaner side of this curve is used to yield the same power with lower fuel consumption. Great care must be taken when attempting this, as combustion temperatures can skyrocket quickly and little margin is left for knock.

For best economy we swing to the other side of the stoichiometric balance with a target of about $\lambda \approx$ 1.05 (about 15.5:1 A/F). Since fuel economy is targeted under cruise conditions, plenty of airflow is usually available. In this case we look to get just enough power to maintain a constant vehicle speed. To achieve this, we mix just enough fuel that, if burned completely, makes enough power to maintain current conditions with enough excess oxygen to ensure that no fuel molecules go unused. This is best accomplished at the peak thermal efficiency point of the engine. To make sure all fuel molecules have sufficient air to react with, some amount of excess air is used. Going too lean merely exhibits a lean misfire condition in which the fuel molecules get so spread out that combustion reaction from one molecule to the next becomes sporadic. Again, leaner mixes for extended cruise periods yield higher oxide of nitrogen emissions and exhaust gas temps in addition to better economy. This can be seriously detrimental to catalyst life and emissions in general. Lean

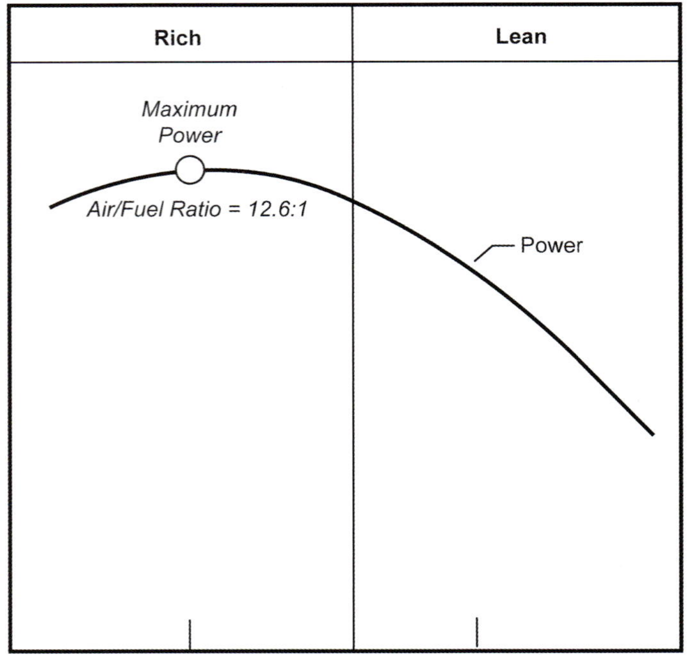

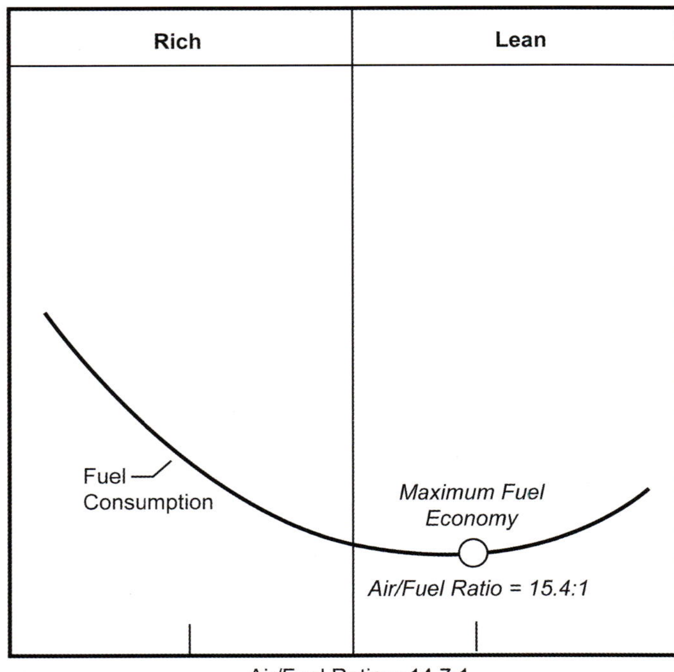

Figure 2-4 *Lambda influence on power output. A slightly rich mixture generates best output. Just how "rich" is best will ultimately depend upon the exact engine combination. (Nate Tovey)*

A slightly lean mixture with yield peak thermal efficiency of the engine and hottest burn temperatures. The exact location of this peak again depends upon specific engine design. (Nate Tovey)

cruise conditions also make for a larger transition back to stoichiometric or richer for acceleration. These reasons add up to making a truly lean cruise more trouble than it is worth to most performance calibrations.

At any given operation point, the fuel requirements of an engine can be described in terms of power. Engineers often refer to Brake Specific Fuel Consumption (BSFC) as the amount of fuel necessary to make the current power. The units for this are self-explanatory: pounds per horsepower-hour. Knowing this number (or at least having a good estimate of it) helps to determine a vehicle's fuel system requirements. That is to say that an engine operating at full throttle and a BFSC of 0.5 lb/hp-hr expecting to make 500 hp can have its fuel requirements calculated as follows:

(500 hp) x (0.5 lb/hp-hr) = 250 lbs/hr of fuel total

If this is an 8-cylinder engine with individual injectors, we can estimate required injector size as:

(250 lbs/hr) / 8 = 31.25 lbs/hr for each injector *minimum*

This is the minimum fuel requirement to support the intended power level. However, this leaves no room for error and doesn't work well in the real world. Fuel injectors operating at 100% capacity usually won't do so for very long. In order to protect against unforeseen conditions (cooler air, more power, richer required mixtures) and give the injectors a little bit of safety margin, it is typical to leave some degree of "cushion." 20% is ideal, but it is possible to work with less. So assuming we only want to use 80% of the injectors' capacity, a new recommended injector size is calculated:

(31.25 lbs/hr) / (0.8) = 39.06 lbs/hr (≈410.9 cc/min) for each injector *safely*

Knowing a required, safe fuel delivery rate of the injectors also points toward a required fuel pump output:

(39.06 lbs/hr) x (8 injectors) = 312.48 lbs/hr total fuel supply

With a known density for gasoline of approximately 6 pounds per gallon, pump flow rate is calculated:

(312.48 lbs/hr) / (6 lbs/gal) =
52.08 gal/hr (≈197.4 L/hr)
desired fuel pump flow rate

Spark Events

Once a cylinder is filled with a mixture of fuel and air and compressed, it holds a tremendous amount of energy. The release of this energy is triggered by the spark event. The spark is generated by closing a circuit on the low (primary) side of the ignition coil. The flow of electrons through the primary circuit of the coil excites electrons in the secondary side of the coil to an even higher voltage. The high voltage looks for a path to ground through the plug wires, distributor, and spark plugs. With a narrow gap and high voltage present, the electricity arcing across the gap has enough energy to reach well over 5,000 degrees F in this narrow gap. This begins a chain reaction of combustion between the compressed air and fuel in the cylinder. How long this arc is present is referred to as "dwell." Dwell should be just long enough to ensure proper ignition of the fuel/air mix without excessively draining the coil saturation.

The entire mixture does not instantaneously change state and composition. Because the burning of the air/fuel mixture takes time to create the dramatic pressure rise, it becomes necessary to start the process slightly ahead of time in order to make sure pressures are at their peak at the right time. This "leading" of the ignition event with respect to TDC is what determines the amount of time before peak pressure is reached. (**Figure 2-5**) Adjusting the ignition lead shifts the arrival of peak combustion pressure relative to piston motion. The object is to get peak combustion pressure just after TDC to maximize usage of the power stroke. If ignition happens too late the piston can literally run away from the expanding mixture. Start ignition too soon and the pressure may peak before the piston reaches TDC. The momentum of the engine pushing against the piston fighting tremendous pressures in the cylinder can lead to piles of expensive broken parts. Adjusting this lead time to deal with changes in operating conditions is the art of calibrating spark advance.

Under most conditions, an engine operates with many different

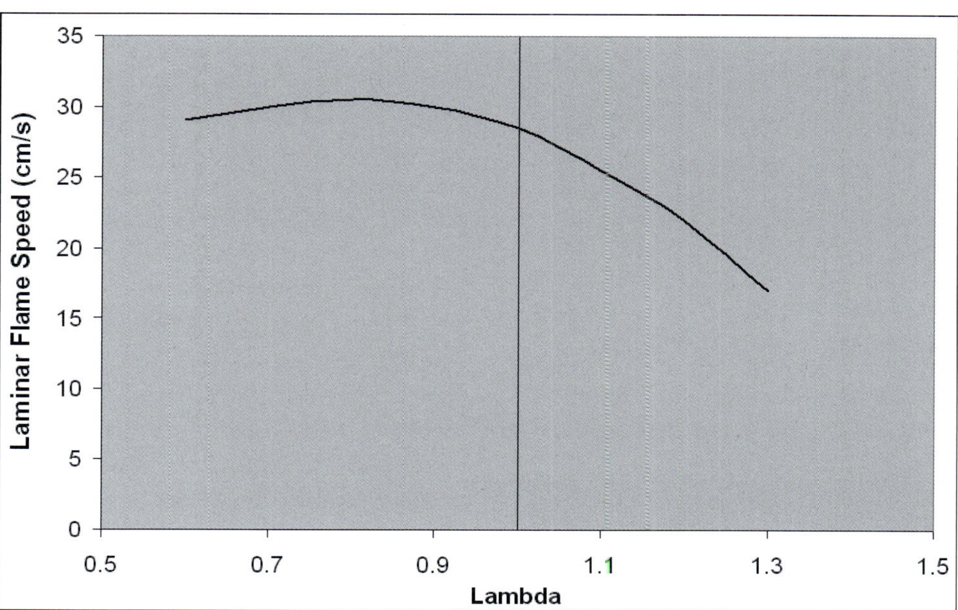

Shown here at 1 atm pressure (14.7 psia), flame speed is proportional to lambda. The trend remains the same at elevated cylinder pressures, although the magnitude of the speed is increased. (Nate Tovey)

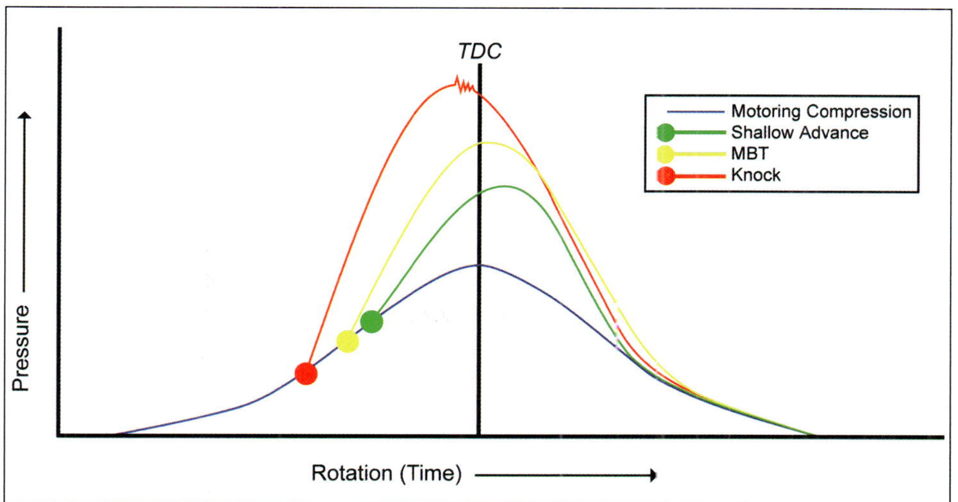

Figure 2-5 *Cylinder pressure versus rotation (time) for various conditions. The blue line represents the static compression of the engine without any power event. The red and yellow lines show pressure rise associated with igniting an air/fuel mixture prior to TDC. (Nate Tovey)*

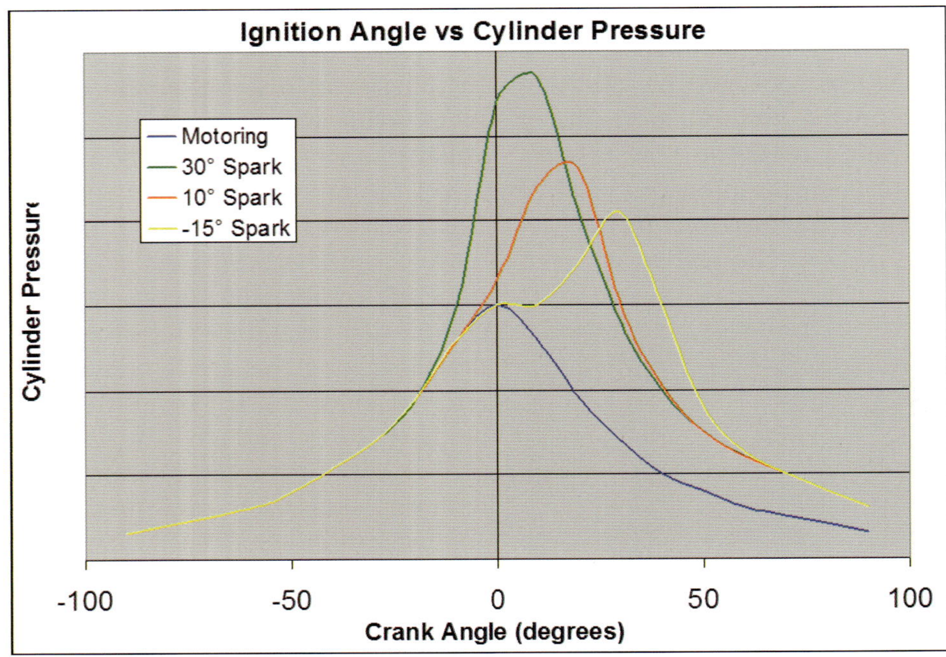

Cylinder pressure effects of various spark advance settings. Notice how cylinder pressure increases with increasing spark advance. (Nate Tovey)

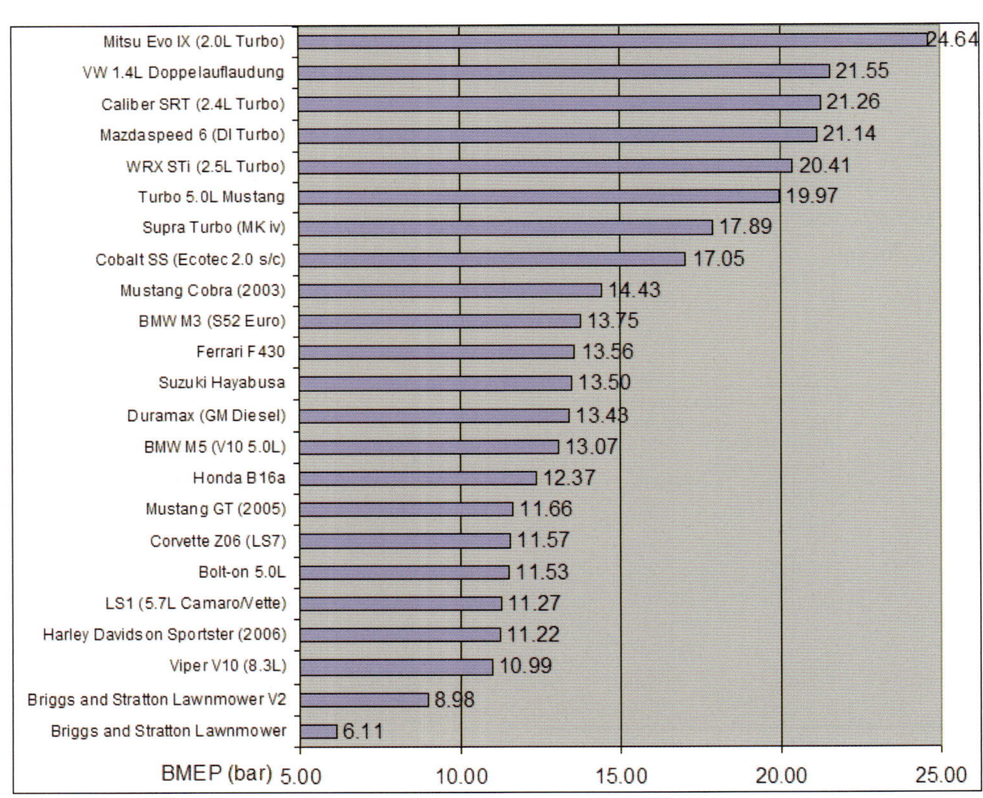

Brake Mean Effective Pressure of various modern engines. Normalizing the engines' torque output against displacement allows for easy comparison of the efficiency of many different engine designs.

spark advance values. As long as knock is not encountered, the range can be fairly wide. Somewhere near the middle of this range is a spark advance value for a specific speed-load condition that results in the most engine torque output. This spark advance value is known as "Maximum Brake Torque" (MBT). Actual MBT timing for an engine changes depending on speed and load. At high engine loads, the onset of knock may prevent operation at this spark value, but the object is to try and run as close to this as possible to get the most efficiency out of the engine without causing damage.

The pressure rise resulting from combustion is the source of engine power. To compare the efficiency of one engine to another or even different calibrations on the same engine, we can look at the pressure generated inside the cylinder. Since it is this pressure that pushes on the piston creating torque at the crankshaft, it becomes a good comparison point. Normalizing this pressure with overall displacement allows engineers to work with a number they call "Brake Mean Effective Pressure" (BMEP). Brake mean effective pressure works out such that it is equal to shaft (brake) work divided by engine displacement. BMEP takes into account all pumping, friction, and accessory losses within an engine. The same number without taking into account these losses is "Indicated Mean Effective Pressure" (IMEP). Indicated mean effective pressure is closer to the raw average combustion pressure and still serves as a good indicator of how much total energy is being released during combustion. Normalizing against engine displacement allows comparisons of relative efficiency between engines of varying sizes.

THE GOOD OL' DAYS

A typical four-barrel carburetor as installed on a single plane manifold. This arrangement is best suited to high RPM operation. (Nate Tovey)

Carburetion

If you can tune a carburetor, you can tune EFI. Every single component of a carburetor has its counterpart in the tables of modern EFI systems. Many die-hard racers swear by their trusty old double pumpers and it's tough to argue with results.

The reasoning is dead simple. Carburetors by design MUST deliver an amount of fuel directly proportional to the velocity of air flowing through the venturis. Let's start with the fuel side. Obviously, a pump is required to transport fuel from the tank to the carburetor, but the side effect is a lightly pressurized feed to the bowls. The bowls of a carburetor act much like a water tower. The higher the level, the more pressure we get—even on a relatively small scale. If we introduce a hole near the bottom, we get more pressure if the level above this hole is higher or if there is added pressure on the liquid itself. Controlling the size of this hole (jet size) controls the flow rate of the fuel once something begins to draw from the bowls.

The venturis of a carburetor act exactly like the wings of an airplane. As the velocity of the air increases in the smaller diameter of the venture, Bernoulli's Law states that the pressure must drop. This yields a low-pressure zone immediately outside the fuel discharge. This pressure differential is what draws fuel from the bowls into the engine. Since both the venturis and the fuel discharge are fixed in size, their flow ratios are also fixed. Very simple: more pressure drop equals more fuel flow.

The biggest compromise here is that, other than at idle where the mixture screws have enough authority to change things, we are locked to one flow ratio of air to fuel for the primary circuit and one flow ratio for the secondary circuit, if so equipped. This means tuning a carburetor often boils down to a compromise of desired air/fuel ratios between cruise, light throttle, and wide open throttle (WOT). With crude adjustments at best to control transition and no real way to adjust

The "dry" intake tract of a modern fuel-injected engine shown from filter to throttle body to manifold. Fuel is not added until just before the air enters the cylinders, allowing for more convoluted path design without puddling. (Nate Tovey)

Above: State of the art, circa 1969. The 427-ci Copo Camaro was the epitome of musclecars of the era, and the inspiration for today's Z06 Corvettes. (Nate Tovey)

Fuel flows from the bowls, through the jets, and out into the venturis of the carburetor. The smaller area of the venturi creates a low-pressure zone that draws fuel into the air stream. (Nate Tovey)

The long individual intake runners can be seen here. The long intake path is tuned to improve cylinder filling efficiency at low engine speeds.

for nonlinear performance across a wide RPM range, there is room for improvement. Additionally, since actual fuel flow is proportional to air velocity in the venturi rather than actual air mass flow, air/fuel ratios can change slightly with changes in ambient conditions. The term "good enough" works for many racecars, but leaves a lot to be desired on a daily driven car expecting good economy and emissions.

Timing

The trusty old distributor is a reliable, but crude method of controlling ignition lead. Assuming that there is little play between the crankshaft and distributor, static ignition phasing is easily controlled allowing for accurate timing of the spark event relative to piston position. However, automotive engines are not static devices. Increases in engine speed mean less actual time between crank angles, so ignition advance must increase just to maintain the same amount of burn time between the first spark event and TDC.

On mechanical distributors, this increasing advance is accomplished with a set of rotating weights secured by springs. Adjusting the mass of these weights and the tension of the springs gives a crude adjustment to the phasing of the distributor's shaft relative to the crankshaft, advancing ignition timing at higher engine speeds.

To compensate for engine load, many mechanical distributors incorporate a secondary adjustment from a vacuum diaphragm. This diaphragm employs an actuator rod that also shifts distributor shaft phasing relative to manifold pressure. Since lower loads exhibit increased vacuum, timing is advanced to compensate and increase performance. As load increases, the diaphragm returns to its static position, reducing timing to prevent detonation from excessive spark lead. This allows maximum WOT timing to be easily set with the vacuum line removed and better fuel economy and power at cruise and idle when vacuum is present. The drawback to the vacuum advance is the lack of adjustment range. Combinations with

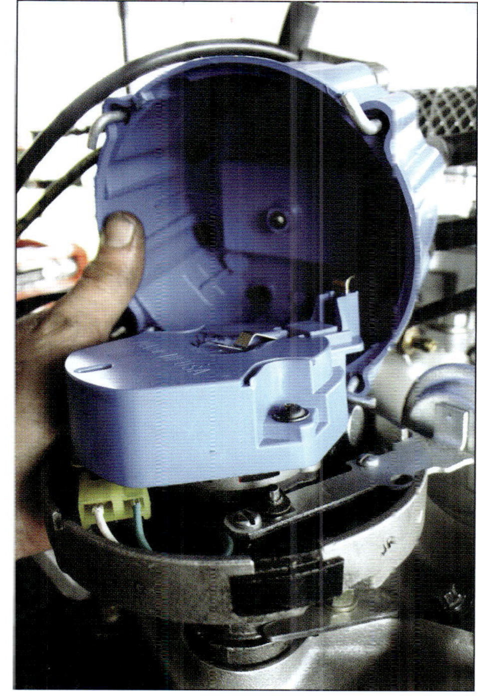

Typical mechanical distributor with vacuum advance. As vacuum increases, the phase of the rotor is adjusted by the arm on the right, increasing ignition advance. (Nate Tovey)

larger camshafts or certain intake manifold designs often render load adjustment based on the single vacuum source unreliable. This often has the tuner resorting to removing the vacuum reference to keep ignition advance more stable.

Ideally, there would be a way to accurately adjust the amount of spark advance versus speed and actual engine load. EFI systems give the calibrator the flexibility to make timing adjustments at more than one or two points.

Cam Timing

Another adjustment that is often available, but admittedly more intrusive and messy, is changing camshaft timing. Advancing or retarding the camshaft relative to crankshaft angle

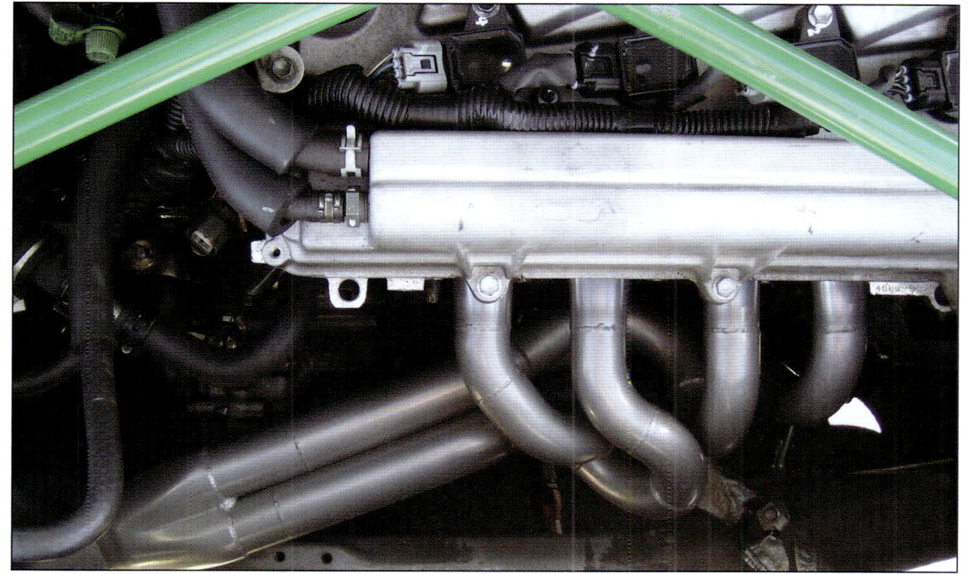

Ram tuning can work on both the intake and exhaust side of an engine. The tuned header on this Toyota MR2 greatly improves exhaust scavenging.

The timing chain links the camshaft (large gear) to the crankshaft (small gear) and determines the relative position. Most older engines do not allow for easy adjustment since the timing chain must be entirely removed to re-index the crank gear. (Nate Tovey)

Most actual calibration work at the OEM level is done at engine dyno cells like the one shown here. Every possible engine parameter is carefully controlled and measured to ensure repeatability of the tests.

can shift the tuning characteristics of the engine to allow greater efficiency at higher or lower speeds, respectively. Advancing the camshaft generally gives better low-end and midrange torque by closing the intake valve sooner, trapping the charge in the cylinder before it gets a chance to revert back into the intake port as the piston begins to ascend. Since most systems exhibit relatively little exhaust scavenging effect at lower speeds, opening the exhaust valve earlier also reduces static pumping losses by venting the high pressure more immediately to the exhaust system.

Likewise, retarding cam timing allows the intake valve to remain open later where the ram tuning effect can continue to pack air into the cylinder at the end of the intake stroke at higher speeds, making more power. As speeds increase and scavenging efficiency increases in the exhaust, larger volumes of gases can be extracted from the cylinder with a later valve closing. This leaves more empty volume for the intake charge to fill on the following cycle.

Getting at the camshaft drive gear usually involves removal of the

The addition of full electronic controls gives the calibrator much greater flexibility in the tuning process. What used to take hours of dirty work can now be accomplished in minutes with just a keyboard and the right software. (Nate Tovey)

front cover and accessory drive, which is usually less than practical during the tuning process on a working automobile. Since it is difficult to adjust cam timing, a compromise needs to be

The blue boxes in the foreground are signal processors for in-cylinder pressure measurements. The business end of a million-dollar production-level engine dyno can be seen in the background.

found that gives the best performance split. If the camshaft lobe positions could be altered with respect to speed and load, greater overall engine efficiency could be had without actually changing components.

TAKING MEASURE

Before we delve into the tables that control engine operation, it is important to understand what the computer sees. The old saying, "garbage in, garbage out," applies directly to EFI systems. More often than not, the cause for a poor-running car lies not in a bad calibration, but rather in a bad calculation. This is to say that engine output controls are formed based on inputs and calculations. A perfectly good calculation on a bad input parameter drives just as poorly as a poorly tuned car. Although not all sensors are absolutely critical to engine function or drivability, there are a handful that the EFI computer can't live without. Likewise, sometimes we have redundant sensors where only one really impacts the final calculation of engine output controls and others are merely monitors.

Throttle Position

The throttle position sensor (TPS) is one of the most critical inputs for any EFI system. Think of it as the volume knob on the stereo in form and function. The TPS tells the computer exactly where the throttle blade is so that it can be determined whether the driver is attempting to idle, cruise at a steady state, accelerate, or decelerate. Additionally, the rate and direction of change of this sensor helps the computer determine if the driver is attempting to change states. Most TPS sensors are basically a rotary potentiometer that varies output based on position around a dial. The further up the dial, the longer the electrical path gets through a set of resistors. We usually read output in a 0 to 5-volt scale with 0.5 to 1.0 v usually indicating closed throttle and 4.5 v or greater indicating wide open (WOT).

It is important to know exactly what the threshold is between closed throttle (C/T) and part throttle (P/T) for calibration to ensure that the computer actually uses the idle tables when the driver's foot is off the pedal. It is a common mistake for a car owner to open the idle screw on the throttle body without checking TPS output. If the blade is opened beyond the C/T threshold to prevent stalling, the TPS must be readjusted to reflect the new closed throttle position. Likewise, a stretched cable or bent linkage may prevent the

Mounted directly to the throttle shaft, the TPS sensor (arrow) reports actual blade angle to the PCM. (Nate Tovey)

computer from seeing WOT even though the blade is 99% open. Since the effective flow area changes so little between 70% and 100% blade opening, many systems consider anything over 70% or so to be wide open depending on RPM.

Coolant Temperature

Temperature of the engine itself is another extremely important monitoring point. This is a two-way street. Engines have a relatively narrow temperature band in which they operate most effectively. Too cold and fuel has trouble atomizing before the combustion process. Too hot and preignition in the chamber has almost identical negative effects to knock or expansion and distortion risk, warping critical sealing surfaces. Actual desired operating temperature depends upon desired engine usage. Most current OEM systems are thermostatically controlled to about 200 degrees F to allow for ideal combustion and emissions. Typically, a 20-degree drop to about 180 degrees F nets cooler chamber temperatures and allows those few extra degrees of spark advance that make more horsepower. Much like the mechanical choke on a carburetor, EFI systems allow for enrichment and added idle speed based at cold temperatures. Going too cool on the thermostat opening temperature can keep many OEM processors in the warm up routine skewing target fuel delivery and idle speed. A skilled calibrator can change the parameters that determine what is "warmed up" to avoid excess enrichment. However, intentionally running a cold engine and head temperature is usually reserved for drag race applications where emissions and cylinder wash from excess raw fuel are not a concern.

Knowing how much enrichment to add depends directly upon actual temperature. This comes from a sensor in either the coolant path or mounted directly in the cylinder head itself. These sensors are typically a basic thermistor with a resistance that varies directly with contact temperature. Most ECT sensors are of the negative temperature coefficient (NTC) type. NTC thermistors reduce in resistance as temperature increases. With a steady input voltage (usually 5 v) to the thermistor, increasing temperatures are read as higher return voltages from the sensor as a result of the dropping resistance thanks to Ohm's Law.

If the sensor is placed in the coolant path, it is important that no air pockets are present. It is a common error to register a relatively cold input signal when the sensor is actually sitting in a steam pocket out of contact with actual engine tempera-

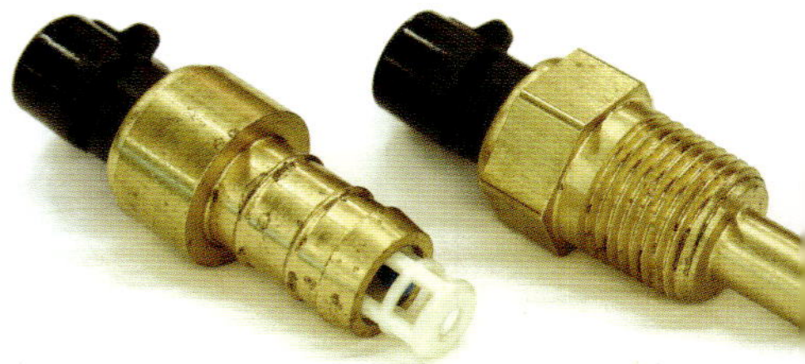

Intake Air (left) and Coolant (right) Temperature sensors are usually NTC thermistors with slightly different housings to accommodate their installation environments. (Nate Tovey)

tures. Since most EFI systems have safeguards in the code to allow for higher idle speed (more coolant circulation), increased electric fan activity, and richer fueling conditions at high engine temperatures, this input can be an engine saver. A cylinder head temperature sensor mounted directly to the casting reduces the chance of this error. The thing to keep in mind here is that cylinder head temperatures are typically 8 to 15 degrees F warmer than coolant temperatures at any given time due to conductivity.

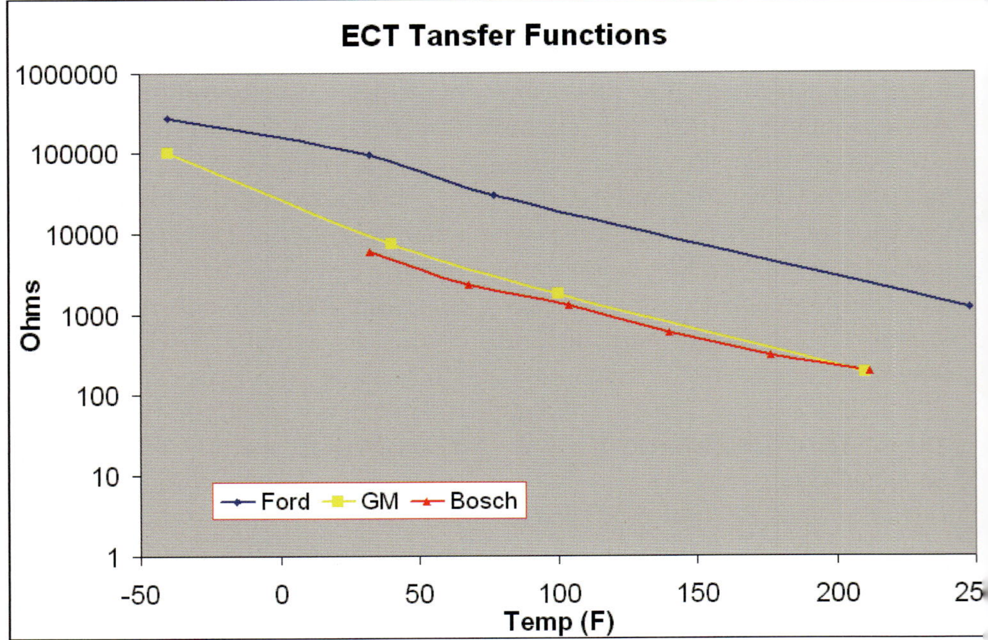

Transfer function of several different ECT sensors. Note the logarithmic scale for resistance that yields large changes with temperature.

Air-Inlet Temperature

Air temperature has a direct effect on density as well as burn rate. As air is heated it gains volume and loses density. In speed-density systems, this measurement is critical to determining exactly how many oxygen molecules are available for combustion in the current cycle. Since the engine displaces the same amount of manifold air in any given cycle, temperature directly affects the density (number of available oxygen molecules) actually making it into the chamber. On mass air measurement systems, air-inlet temperature does not directly adjust calculated charge fill for fuel calculations, but it is still used as a modifier to control

The IAT sensor (arrow) should always be installed in a location that best represents the temperature of the air entering the cylinders. In the case of a supercharger, this means placing it after the compressor to measure any heat soak as seen on this turbo Toyota Supra engine. (Nate Tovey)

burn characteristics. Inlet temperature is directly proportional to allowable spark advance assuming constant fuel delivery. Colder inlet temperatures mean more timing advance can be used, resulting in more available horsepower. Conversely, much hotter inlet temperatures (often a result of supercharging) require reduced timing to avoid the knock that comes from the increased burn speed.

Most inlet-temperature sensors are very similar in design and function to NTC coolant-temperature sensors. Since air density is so much lower than coolant density, sensor elements can be unshielded without risk of damage, making for faster response to changes in conditions. It is important for the calibrator to understand where the inlet temperature is being monitored on forced induction applications. While some OEM supercharged applications actually employ two inlet-temperature sensors to monitor both ambient conditions and actual inlet-port temperature, most systems only have one such input. Ideally, these sensors should be placed as near to the cylinder-head intake port as possible so that the computer sees actual charge temperature. This allows for tighter control in supercharged applications where charge temperatures vary from 90 degrees F ambient to over 300 degrees F in non-intercooled vehicles. Again, a skilled calibrator can compensate in the tune for a supercharged engine with an inlet-temperature sensor installed ahead of the compressor with a shift in the temperature compensation function. By shifting this function and leaving an appropriate safety margin, everything can work just fine, but it's easier to relocate the sensor for a more accurate signal.

Manifold-Surface Temperature

Some systems add another sensor to monitor manifold-surface temperature. This allows for an additional routine in the computer to model heat transfer from the intake manifold to or from the intake charge immediately ahead of the cylinder. Again, this effect is more pronounced on speed-density systems where the actual air mass entering the cylinder must be calculated and temperature plays a bigger part in required fuel delivery. These sensors are usually almost identical in design and construction to the NTC coolant-temperature sensors

Mass Air Flow

Since internal combustion engines are so sensitive to air/fuel ratio, it is important to know exactly how much of each component is entering the engine at any given time. The best way to ensure accurate fuel delivery is to have accurate air measurement before calculating anything. Speed-density systems are constantly making calculations of estimated airflow based on pressure, temperature, and engine speed. Mass air systems employ a sensor that directly measures air mass flow into the engine. Basically, the more air molecules moving past the sensor, the more the signal changes.

Early units used spring-loaded doors that were pushed further open by the force of the incoming air. The density and velocity of the incoming air is proportional to the force on the door. The doors of these sensors were then attached to a rotary potentiometer much like the throttle position sensor. The drawback to these units is the lack of temperature compensa-

---

tion. These early sensors require further calculation based on inlet temperature to determine the actual air mass.

Modern units use a small heated wire or film element suspended in the air stream. As more air molecules pass by the heated element, more current must pass through it to maintain a constant temperature. As current increases in the heated element, the output of the sensor changes. This heated element is cooled in proportion to the mass of air flowing over it, incorporating temperature compensation for a direct mass flow input to the PCM. For Ford vehicles, output is a 0 to 5 v signal. For GM vehicles, output is a 0 to 12,000 Hz wave. Either way, the transfer function ends up as an exponential curve with more resolution at low flow to accurately meter idle and cruise loads. It takes relatively large increases in airflow at the high end to make a change. Although the wire element itself is relatively small, it is placed in the meter housing such that the amount of air crossing the element is directly proportional to the total amount of air entering the engine. Most OEM mass airflow sensors are sized to be slightly smaller than the rest of the inlet plumbing in an effort to control velocity and increase accuracy.

A closer view of the MAF sensor shows the heated wire element that hangs into the air stream. The cooling rate of this wire is directly proportional to the mass of air moving past it. (Nate Tovey)

Vane Air Meter as installed in a 1990 BMW 3-series. Notice the curved balance chamber for the flapper door. This design still required significant temperature compensation to work properly. (Nate Tovey)

A common mistake made by many performance enthusiasts is to cut out a portion of the mass airflow (MAF) sensor to improve total flow. While total flow is increased by doing this, the side effect is a change in the ratio of air across the metering element to total flow, reducing the output of the MAF sensor at any given actual flow rate. The net effect of this is a leaner engine-operation condition resulting from fuel calculations based on a lower airflow input to the computer. For example, cutting the center divider out of a GM LS1 MAF sensor typically shifts the output down by about 7%. The bigger concern is that this modification does not simply shift output down across the range, as low speed flow demonstrates a more pronounced effect. While many OEM systems run a safely rich

The modern hot wire element MAF sensor is an extremely accurate method of measuring incoming air mass. Notice the honeycomb flow element and wire screen that help reduce turbulence entering the meter. (Nate Tovey)

The airbox on the Subaru WRX STI features an integrated MAF sensor. This particular unit is made by Denso and is a hot film element.

air/fuel ratio at WOT, leaning out the mix by changing MAF output can lead to knock.

The problem that many performance enthusiasts encounter is that the OEMs have not intended the power (and total airflow) of the vehicle to be drastically increased. Most OEM mass airflow sensors are scaled to have the most resolution possible within the anticipated range of possible flow requirements. As airflow increases with increased power production, we often find that this anticipated maximum flow can be exceeded. The result is an OEM meter that reaches its maximum output before the engine peaks. This is often referred to as a "pegged MAF sensor." Due to the limited measurement range and a typically small size, it is common to replace the OEM MAF sensor on performance vehicles. Aftermarket MAF sensors invariably have a different output

from the original, even if only slightly. The ability to shift this output upward significantly makes their use almost mandatory on higher output engines. Once the OEM

meter has been replaced, it is up to the calibrator to change the tables in the computer to reflect the new flow rates at each output point. If this new output curve is modeled correctly, the result is accurate measurement even at higher flow rates. As long as the computer knows exactly how much air mass is entering the engine, calculating the proper amount of fuel is relatively easy.

Aside from sensor flow and output range, the calibrator must also pay attention to the actual installation of the MAF sensor. Since we are ultimately dealing with airflow through a confined path and only metering a small portion of it to determine total flow, it is important to understand the effects of plumbing on meter outputs. The primary concern is the assumption of laminar flow across the area of the MAF sensor. If airflow across the sensor is greater on one side than another, the output can be skewed relative to the clocking of the metering element. The common source for such a problem is usually a bend in the

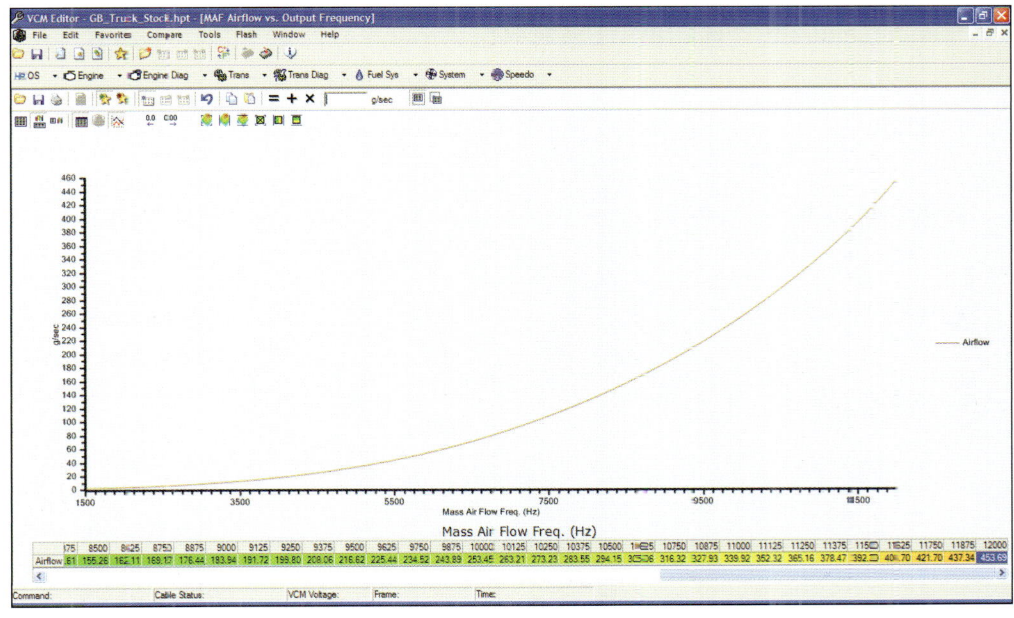

Typical MAF sensor output. Notice the exponential shape of this curve with large changes in output frequency at low flow rates.

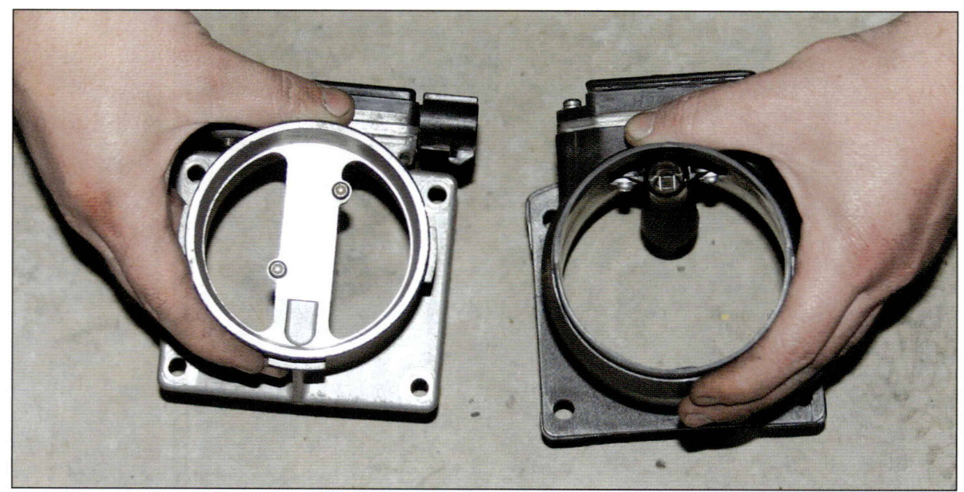

A ported MAF sensor disrupts the smooth output of a factory sensor. Using a MAF that has been modified like the one on the right requires careful adjustment to the transfer function in the PCM to correct for its now unique output curve. (Nate Tovey)

intake plumbing immediately ahead of the MAF sensor. Since airflow is biased toward the outside radius of a bend, placing the metering element on the outside radius of a bend yields an output that indicates higher total flow than is currently present. The opposite is also true with reduced outputs resulting from element placement on the inside radius of a bend. Generally speaking, if the MAF must be placed following a bend in the inlet plumbing, it should be done such that the element is perpendicular to the plane of the bend. Air filter design or placement in the vehicle may also demonstrate a similar condition.

Constant diameter approaching the MAF sensor is also desirable. Remember that the same mass flow through a smaller diameter increases velocity. If the diameter of the inlet plumbing increases immediately ahead of the MAF sensor, it often results in an unstable charge flow with plenty of tumbling. This tumbling air may cause the wire element to read incorrectly depending on inlet flow and velocity. A length of straight, constant diameter pipe is preferred before any MAF sensor. Typically this length should be at least double the diameter. This gives any tumbling flow an opportunity to stabilize before entering the sensor.

Many OEM applications reduce clocking and velocity effects by integrating a laminar flow element into the MAF sensor assembly. The design of these flow straighteners ranges from simple wire screens to honeycombs with an exceptionally high Reynolds number. Many performance enthusiasts looking for extra power mistake the laminar flow element for a genuine restriction and remove them. Benefits of removal rarely justify the loss in metering accuracy and usually fall well below statistical accuracy of most chassis dynamometers. Typical

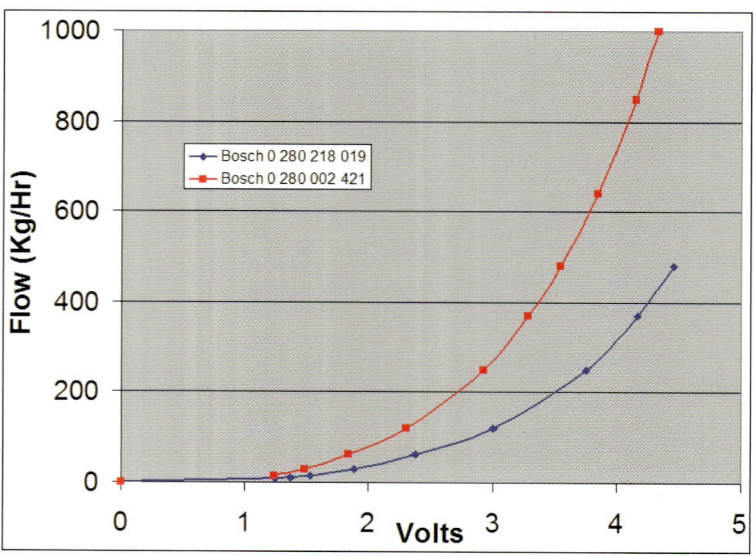

Comparison of output curves for two different MAF sensors. Notice how one sensor requires significantly more airflow to reach the same output voltage at the upper end. The red curve is a meter that can support much more airflow before "pegging."

The wire mesh screen between the airbox and measuring element helps to promote laminar airflow and accurate readings. (Nate Tovey)

The sharp bend immediately in front of the MAF sensor in this photo has a pronounced effect upon readings. This installation often requires an adjustment to the transfer function in the PCM like the ported meter shown earlier. (Nate Tovey)

This proper installation of a MAF sensor in a blow-though setup shows a section of straight, constant diameter pipe leading into the MAF element.

changes I have seen from such modification rarely exceed 4 hp at the wheels even on modified V-8 engines. If a vehicle has a modified inlet system and erratic MAF sensor output, it is often helpful to install some sort of flow straightening. This can be done with a simple wire mesh or length of straight pipe leading into the MAF sensor.

Clockwise from left: MAP sensors from a Dodge Caravan, an aftermarket ECU (GM one bar), and LS1 Camaro. The vacuum connector is clearly visible on each. (Nate Tovey)

Manifold Pressure

The manifold absolute pressure (MAP) is a key element in speed-density systems. Charge density can be calculated by measuring the exact pressure in the intake manifold and knowing an accurate temperature. Air mass can be calculated accurately by knowing the system efficiency at a given point and multiplying that by the density and displacement. This gives the computer a number to work from for airflow and allow for proper fuel delivery calculations.

The manifold pressure itself is measured by a diaphragm that is directly acted upon by the air in the manifold, sometimes connected by a

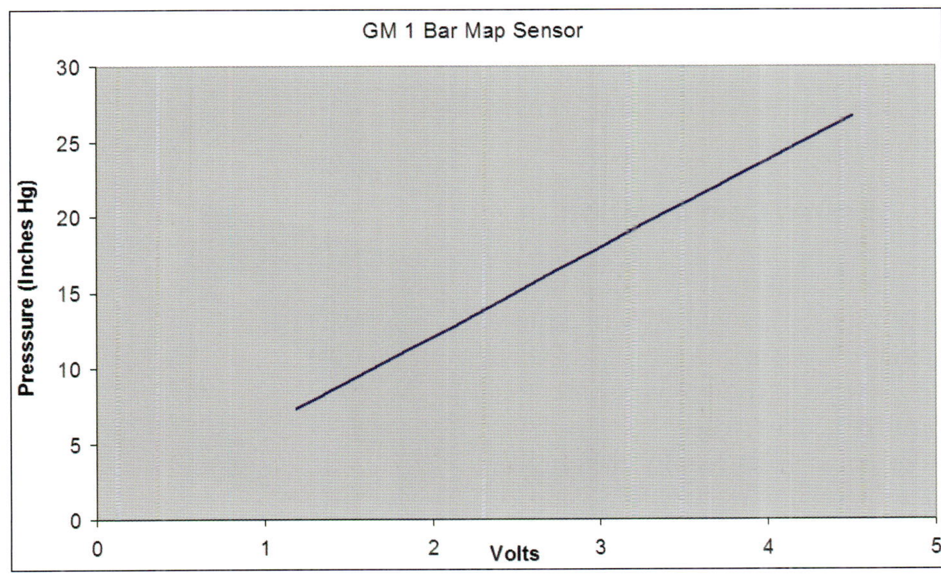

MAP transfer functions for a typical one-bar sensor. Notice the linear output with respect to pressure. (Nate Tovey)

vacuum line. This diaphragm in turn acts directly upon a normal strain gauge. The range of these sensors is usually measured in bar or single atmospheres. That is, a one-bar MAP sensor has a range from full vacuum to one atmosphere of pressure and is used in most naturally aspirated engines. A two-bar MAP sensor has a range from full vacuum to one bar above atmospheric or 14.7 psi of boost and so on. Manifold pressure is directly proportional to engine load, so it is used at the y-axis versus speed for the fuel and spark maps in most speed-density systems.

Barometric Pressure

Barometric absolute pressure (BAP) sensor design and function is nearly identical to that of MAP sensors. The difference is that a BAP sensor is not connected to the intake manifold, but open to the air. The primary function of the BAP sensor is to help mass-air-based systems detect changes in altitude that require adjustments to calculated load and timing. Many modern mass air systems infer the barometric pressure based on MAF sensor output at startup.

Crank/Cam Position

The crankshaft or cam position (CKP) sensor is typically a two-piece system. Part one is a wheel installed on either the crankshaft or camshaft. This wheel has teeth or ridges made of a readily conductive metal, usually steel. Often, one or more of these teeth is offset. Part two of the sensor is a magnetic pickup (Hall effect sensor) designed to provide an output that fluctuates

The camshaft (left) and crankshaft (right) position sensors are similar in design and construction. The difference in length is usually an installation position requirement. (Nate Tovey)

with the passing of each tooth or ridge on the signal wheel.

The CKP sensor performs two primary functions. First and most important is to report instantaneous engine speed. The second function, more critical to sequential systems, is to indicate actual rotational position. The location of the missing or extra tooth on the wheel allows the computer to synchronize the first fuel pulse and spark event with the number one cylinder. Additionally, if the computer fails to see the expected speed or acceleration from the CKP, it can recognize a misfire event. Knowing exactly how far past the last synchronization the misfire happens also correlates to a particular cylinder. This makes it easier for the computer to self-diagnose a bad individual plug, coil, or injector.

Rail Pressure

A less common sensor is the fuel rail delta pressure (FRP) sensor. Injector flow rate increases as pressure across the injector increases. This must be modeled to correctly

The fuel pressure sensor (arrow) on this Mustang is mounted directly to the rail. Notice the attached vacuum line that allows the sensor to measure actual pressure drop between the rail and manifold regardless of boost or vacuum. (Nate Tovey)

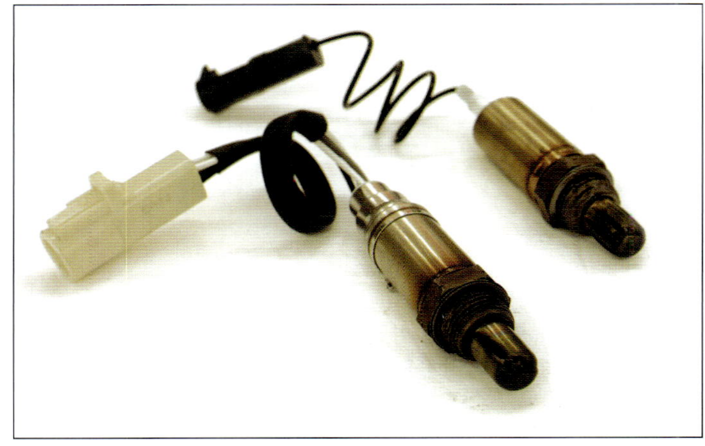

Exhaust oxygen sensors shown with (foreground) and without internal heaters. The addition of the heating elements allows for accurate measurement a shorter time after engine startup. (Nate Tovey)

calculate the amount of fuel actually making it into the intake port. Return fuel systems use a fixed ratio of fuel to intake pressure controlled by a mechanical regulator. In a returnless system, regulation is performed by varying the duty cycle of the fuel pump itself. This eliminates the need for a return line from the engine to the tank. It also eliminates the associated heat in the fuel and splashing upon return to the tank, and reduces evaporative emissions. The FRP sensor is used in returnless fuel systems to provide feedback to the computer for fuel pump control. Since returnless systems use a variable output fuel pump, the computer needs to know how much pressure is available across the injectors at any given time. The design and function of the FRP sensor is very similar to the MAP sensor. The difference here is that the FRP sensor is exposed to fuel under pressure from the rail on one side of its diaphragm and intake manifold pressure on the other. This means that only the difference (delta) between the two effects a change in the diaphragm's position and signal output. For example, a supercharged engine with commanded rail pressure of 40 psi should show 50 psi on a common pressure sensor under 10 psi of engine boost, but a delta of 40 psi from the FRP sensor to maintain constant injector potential.

System Voltage

Although there is never an external sensor for voltage, all EFI systems have means of monitoring it. Power is required to operate the computer in the first place, so it is simple to monitor system voltage directly off the board. This voltage is used to model injector performance as well as ignition coil saturation and dwell time. Lower system voltages require more time for the field to build up in the transformer of the ignition coil as well as the electromagnet of the fuel injector.

Oxygen Sensors

Feedback control systems are as old as electronic fuel injection. Modern EFI systems have the ability to constantly correct for errors between desired and delivered air/fuel ratios. This is done by monitoring the oxygen content of the exhaust gases. Since combustion is just the chemical mixing of oxygen and hydrocarbon fuel, variances in the mix leave an inconsistency in the exiting molecules of the exhaust. An oxygen sensor (lambda sensor) is basically a "battery" that changes potential in the presence of oxygen. A layer of zirconium oxide (ZrO_2) changes the sensor's output voltage based on the partial pressures of the oxygen molecules. The more oxygen present, the lower the output voltage is. The computer uses this output to trim the commanded fuel delivery in what is called closed loop operation. This is where actual commanded injector pulsewidth is actively being adjusted based on feedback from the oxygen sensors. Although this correction is accurate under normal operating conditions, a cold sensor and relatively cold exhaust gases found at startup make for poor accuracy. This is why the computer typically ignores the sensor output for a brief period of time after start, allowing the sensor to reach operating temperature.

To improve accuracy of the measurement process, engineers discovered heat is a vital component. This is the reason oxygen sensors are currently installed as close to the cylinder head as possible and now include heating elements of their own to bring them up to operating temperature as quickly as possible after startup. These heated exhaust gas oxygen sensors are abbreviated as HEGOs. The time between startup and warmed up closed loop operation is critical to emissions since engines typically run slightly rich when cold to prevent stalling. Any reduction in this time period results in lower total hydrocarbon (unburned fuel) emissions.

Most OEM systems employ a standard HEGO sensor on each bank of cylinders with exceptional accuracy near stoichiometric combustion.

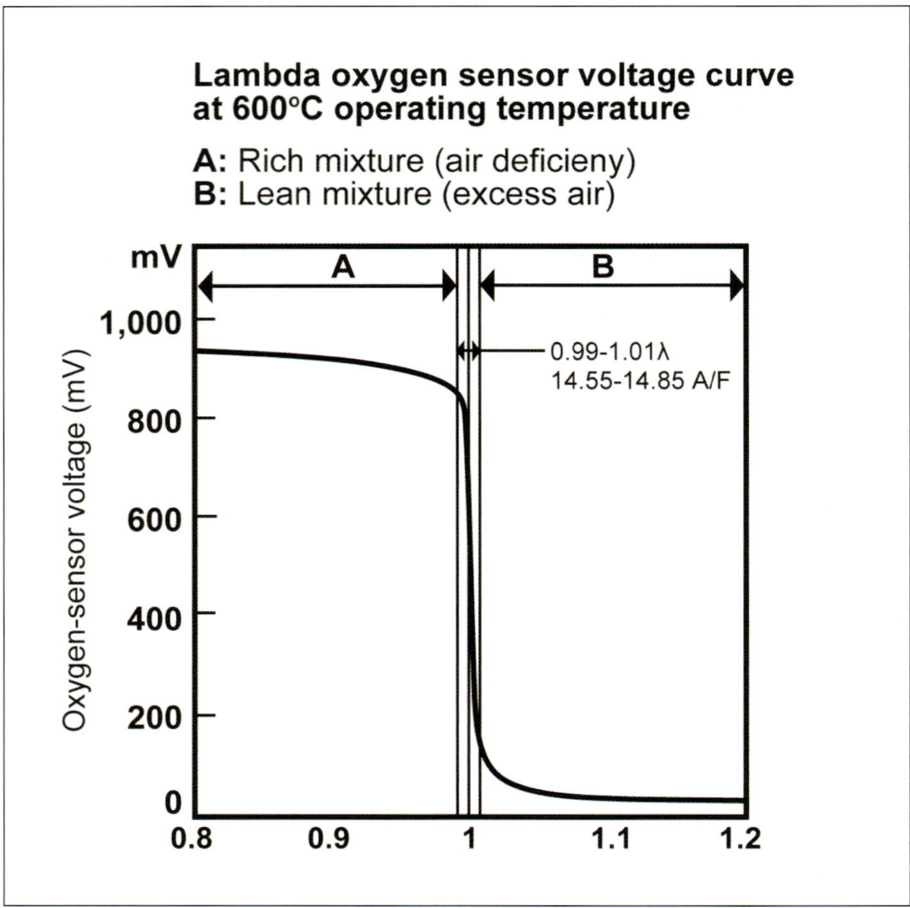

HEGO output versus lambda. The narrow, steep band of switching range near lambda = 1 gives it the nickname "Binary O2 Sensor." (Nate Tovey)

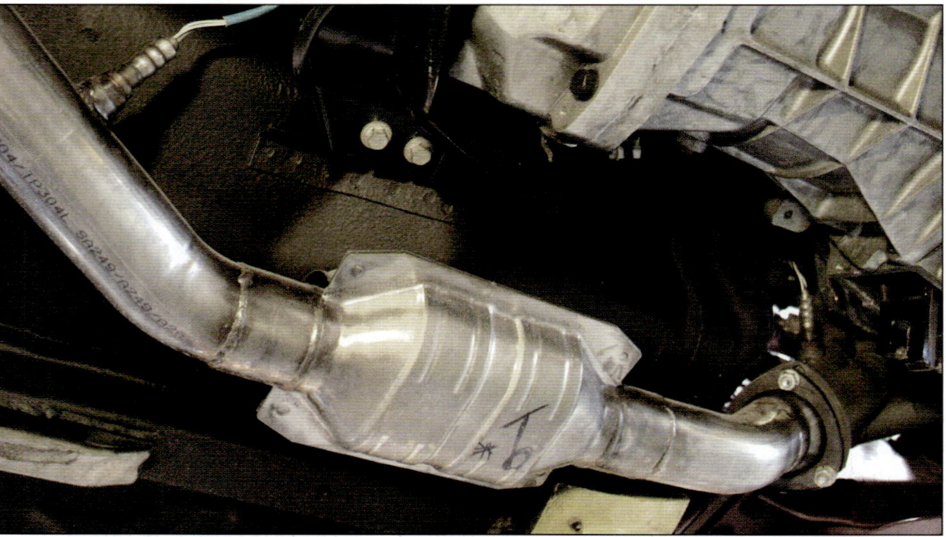

Modern EFI systems employ two HEGO sensors per engine bank. The first monitors engine performance while the second monitors catalyst efficiency for diagnostic functions. (Nate Tovey)

These narrow band sensors are only accurate +/- about one air/fuel ratio from stoichiometric. This means that 13.0 may look "rich" to the HEGO, even though it is too lean to support the intended power level. This is why most EFI systems switch to open loop operation at WOT. It is a common error for inexperienced tuners to rely upon the factory narrow band HEGO when tuning for fuel monitoring under load. I have seen dozens of broken engines directly resulting from what the tuner thought was a rich condition when in fact it was merely richer than the accuracy range of the sensor.

Also important to observe is the effect of leaks and cam timing on HEGO readings. Any exhaust leak upstream or in the vicinity of the HEGO skews the readings lean. Since there are pressure pulses in the exhaust, any leak allows exhaust gases to pulse out and fresh air (containing plenty of oxygen) to pulse in. The result is an inordinate number of oxygen molecules passing by the HEGO, leading the computer to respond with longer injector pulses to compensate for what it thinks is a lean engine. This can lead to a circle of problems where the normally operating engine becomes over-fueled, fouling the spark plugs. This in turn leads to a rich misfire condition sending more unburned oxygen past the sensor and the computer, again trimming the engine richer on the next cycle.

It is almost impossible to accurately calibrate an engine with exhaust leaks, since neither a standard HEGO nor a wideband yields correct results following such a leak. The solution here is to fix the mechanical problem before attempting to tune.

Wideband lambda sensors like this one from Innovate allow the calibrator to know the actual fuel delivery with increased precision. (Nate Tovey)

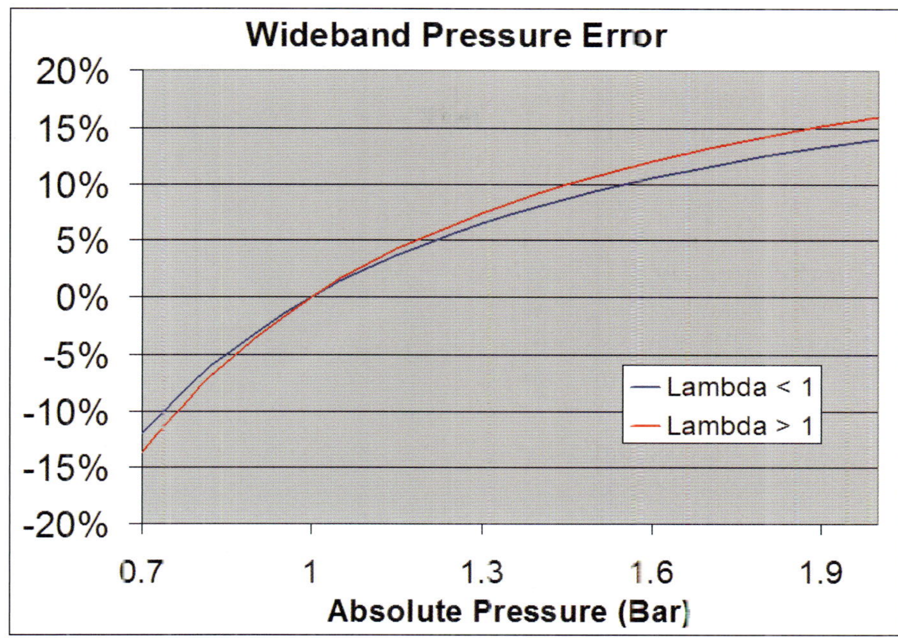

Exhaust pressure can skew the readings of a wideband lambda meter significantly. It's best to avoid this by installing the sensor downstream of a turbo if so equipped. (Nate Tovey)

Likewise, a misfire event where the spark is being blown out before ignition also demonstrates a lean reading on the oxygen sensor output. The solution here is to increase ignition energy or shorten the plug gap to keep the spark kernel alive long enough to ignite the charge.

Cam timing can also cause strange outputs from the HEGO. Although increased overall cam duration adds power compared to stock camshafts, the added overlap can wreak havoc on HEGO control. The wash of unburned intake charge into the exhaust has the same effect as an exhaust leak. This effect usually diminishes at higher speeds and loads since ram tuning tends to feed more of the charge into the cylinder rather than out the exhaust valve. Depending on the amount of overlap in a camshaft, it may be necessary to ignore HEGO outputs at low load or even altogether for fuel trimming purposes.

Many aftermarket EFI systems incorporate more exotic wideband oxygen (WBO2) sensors with an accurate range between 8.0 and 30.0:1 air/fuel. This means that it becomes possible for the computer to actively correct fuel delivery at target ratios well outside the normal range as well as record them for later

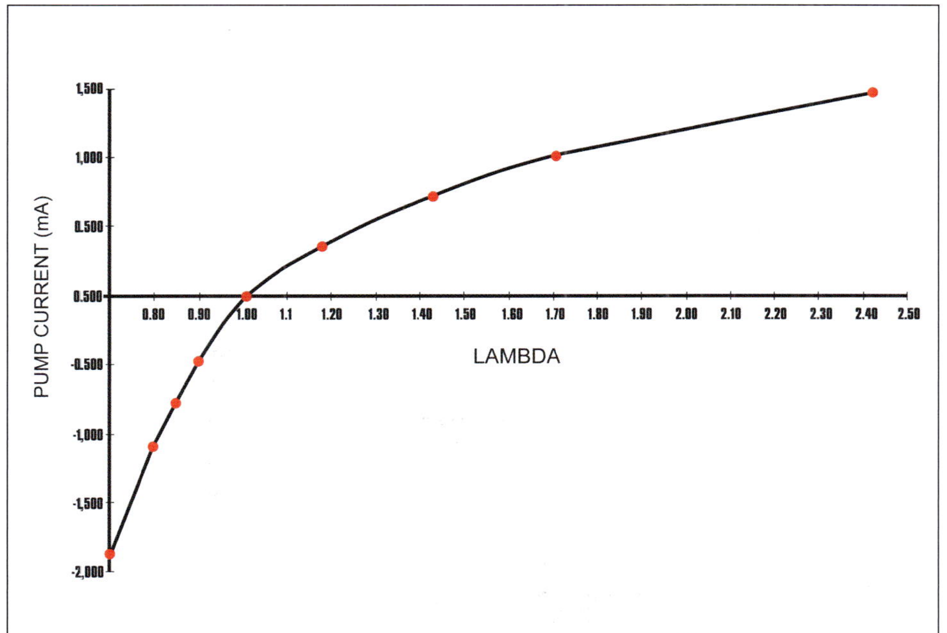

The output of the Bosch LSU 4.x wideband lambda sensor is more consistent over a wider range of lambda conditions. This flatter response curve gives it the nickname "Linear O2 Sensor." (Nate Tovey)

inspection. With improved accuracy over such varied conditions, the wideband oxygen sensor is a tremendously powerful tuning tool for any calibrator even if it is not integrated with the engine controls themselves. Just like with HEGOs, the wideband oxygen sensor must be kept hot to operate correctly and its output must be normalized with temperature for accuracy. Many less expensive stand-alone wideband sensors lack this temperature compensation circuit and pay for it in accuracy.

Exhaust backpressure can also be a concern for wideband oxygen sensor accuracy. Most wideband oxygen sensors are rated to operate between 0.8 and 1.3 bar with reasonable accuracy. Error increases when operated outside of the normal pressure range. There is no sensitivity to pressure at stoichiometric conditions. The pressure sensitivity becomes greater the further from stoichiometric the engine is operated. Increases in pressure make the sensor read further from stoichiometric (i.e., rich reads richer and lean reads leaner). For this reason the WBO2 should be installed after the turbocharger to improve accuracy at high load. Moving the sensor further away from the engine is not desirable for transient accuracy at lower speeds due to delay time, but the accuracy improvement during critical WOT fuel measurement should outweigh this. If high exhaust backpressure at the WBO2 installation point is unavoidable, an extra safety margin should be used on commanded fuel enrichment to compensate for this measurement error effect.

Knock Sensors

Some PCMs are equipped with special sensors to detect the presence

The proper installation of a wideband/UEGO sensor (arrow) in a turbocharged application. Pressure is much lower on the downstream side of the exhaust wheel, which improves accuracy. (Nate Tovey)

of detonation. These knock sensors are basically a microphone with a bandpass filter to isolate the specific range of frequencies associated with knock. The microphone is attached to the engine block or head in the water jacket to take advantage of the faster speed of sound compared to airborne monitoring. Some systems are configured with a range as wide as 4 to 20 kHz, but many are calibrated to be more discriminating based on actual engine characteristics. Most engine knock occurs in the 10 to 12 kHz range. Even if the vehicle is not equipped with such a sensor, some calibrators choose to add their own monitoring circuit during the tuning process to aid in finding the actual detonation threshold of the engine.

In some instances, the changes to the valvetrain, pistons, or exhaust components can inadvertently trigger knock sensor output. GM ran into this problem with the LT4 Corvette engine and countered with a recalibrated sensor. Hard contact between the powertrain (including the exhaust system) and chassis can also be conducted back to the knock sensor, triggering a "false" knock signal. The desired solution is to fix the interference condition rather than desensitize the knock circuit. In some cases where valvetrain or piston noise cannot be avoided, the last resort is to disable the knock circuit only after careful spark calibration to avoid genuine knock that would now go undetected by the disabled sensors.

OUTPUTS

Now that the inputs to the computer have been covered, it's time to look at what the outputs can do. Not that many outputs are required to operate most engines. The advantage of EFI is the ability to very precisely control each of these outputs. Think of jet changes in a carburetor as strokes with an axe and injector control as using a razor blade. Obviously, creating fine details in a sculpture is easier with a razor blade.

Fuel Injectors

The number one priority and function of any injection system is to properly mix the right amount of fuel with the incoming air. This must also be done in such a manner that the fuel is ready to burn as soon as the spark event happens. In a carburetor, the suction of the low-pressure airflow through the venturi draws liquid fuel through a fixed orifice resulting in a misting or suspension of fuel molecule in the incoming air charge in the intake manifold. The manifold design needed to accommodate this "wet" mix features smooth flow paths with gentle bends designed to avoid puddling or uneven fuel distribution.

The fuel injector is basically an electromagnet driven stopper valve for fuel. High-pressure fuel is fed to the top of the injector from the rails. Once the injector is actuated, the force from the electromagnet pulls a pintle upward, opening a small hole to allow the fuel to flow into the manifold. Injectors can use a much higher operating pressure than a carburetor because the spring-loaded pintle design does not leak under pressure. This added pressure also provides for a

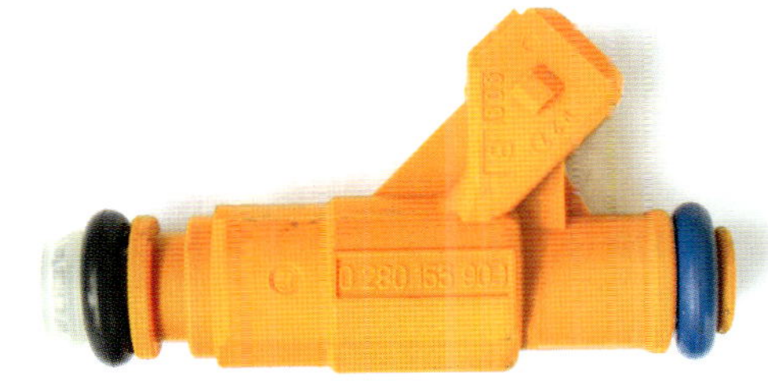

The fuel injector is the key to precise delivery in EFI systems. Siemens manufactures over eight million injectors a year in their Virginia plant, and every single one is flow tested before shipping. (Nate Tovey)

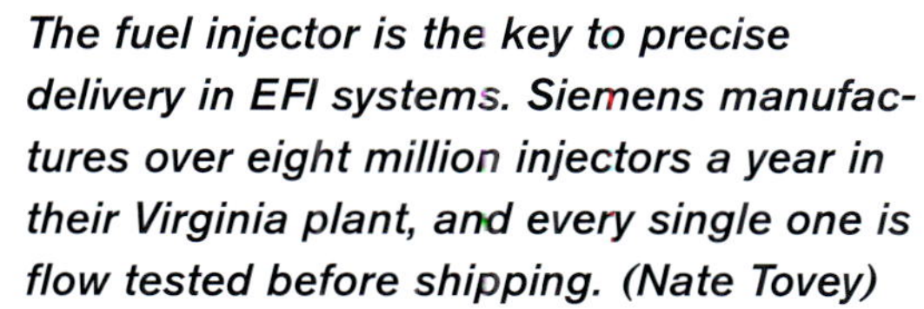

Injectors from different manufacturers shown side by side. The general construction and design is the same despite different lengths and internal flow rates. (Nate Tovey)

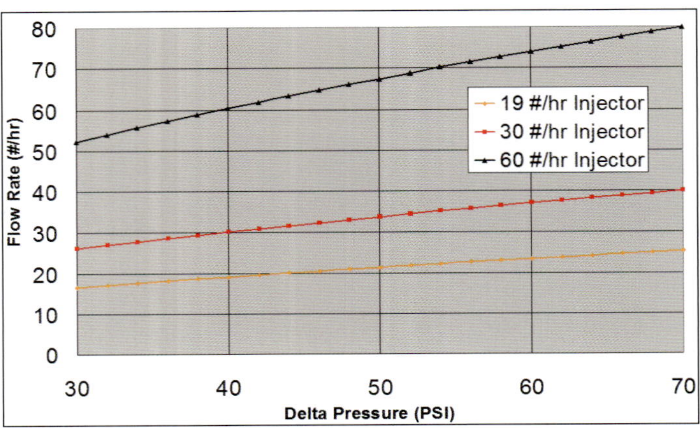

Increasing fuel pressure can be used to provide a higher flow rate out of the same injector. (Nate Tovey)

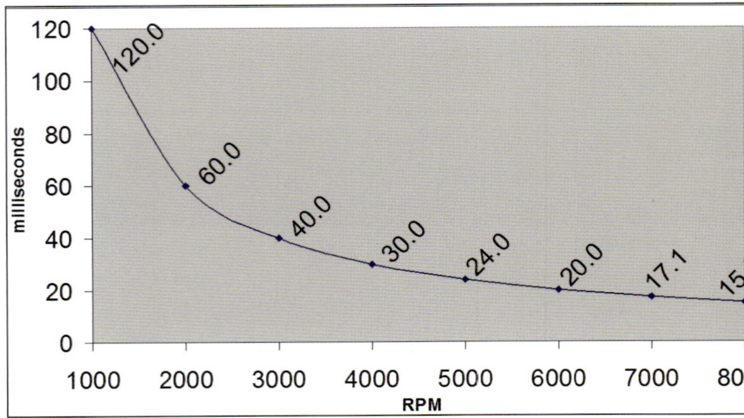

The amount of time available to complete an engine cycle decreases with speed. If the desired injection time becomes longer than this cycle time, the injector will be held static. (Nate Tovey)

more aggressive pressure drop as the fuel enters the manifold, making a finer mist of fuel. The smaller droplet size in turn means better ignition since each individual fuel molecule has more surface area contact with the air charge to aid evaporation and combustion. Additionally, the relatively small size of fuel injectors means that they can be located much closer to the intake valve. This frees up manifold design, allowing for longer runners and tighter, more complex turns to fit a smaller package without puddling. When each individual cylinder can have its own fuel injector, fuel distribution can also be optimized.

Not all fuel injectors are created equal. Injectors are typically classified by the resistance of their internal coil as either "high impedance" or "low impedance." High-impedance injectors are the most common and are used in almost every OEM application today. When measured with an ohmmeter, high-impedance injectors have 10 to 16 ohms resistance. Low-impedance injectors are less common in OEM applications, but readily available in the performance aftermarket. Low impedance injectors typically measure 2 to 6 ohms resistance. It is important to recognize that in most OEM applications, it is not advisable to replace high-impedance injectors with low-impedance injectors. The lower impedance forces the circuit to draw more current through the injector controller, which is really just a transistor. Excess current drawn through these controller circuits often literally cooks them, leading to failure of the injector driver circuit on the board. Since replacement injector driver circuits are so scarce, the result is usually the purchase of a new replacement PCM.

The next division is injector "size," which is really a misnomer for flow rate rather than physical dimensions. Most injectors are rated in either lb/hr or cc/min. These flow rates are actually the static maximum flow through the injector if it were activated continuously. Keep in mind that if the injector is only open for a very short period of time, a correspondingly smaller amount of fuel can be delivered. The length of time that the injector is open during each cycle is known as "pulsewidth." Pulsewidth is usually measured in milliseconds. The ratio of pulsewidth to total available time for that injection cycle is known as duty cycle.

Duty Cycle =

$$\frac{\text{(Pulsewidth)}}{\text{(Total Time Between Ignition Events)}}$$

It is impossible to have duty cycles exceed 100%. If the commanded pulsewidth equals actual cycle time at a given engine speed, the injector is not allowed any time to close. This is known as "static flow," as the rate of fuel delivery is no longer changing. This situation typically happens when injectors that cannot support the necessary fuel flow rate for the engine's power level and fuel consumption rate (BFSC) are used. Uncorrected, this leads to an enleanment of the air/fuel mix. Care must be taken to ensure that this enleanment does not lead to detonation and engine component failure. If more fuel delivery is required, the only available remedies are to either change to injectors with higher flow rate units or increase the fuel rail pressure to artificially increase the flow of the existing injectors.

Fuel rail pressure changes to flow rate are governed by the equation:

$$\text{Actual Flow Rate} = (\text{Injector Flow Rate}) \times \sqrt{\frac{(\text{Actual Effective Rail Pressure})}{(\text{Rated Rail Pressure})}}$$

Keep in mind that changes in manifold pressure also have an effect on actual injector flow rate. If manifold pressure increases from the use of a supercharger, rail pressure must increase 1:1 with boost to ensure that the injector does not become effectively smaller. It is possible to tune around changes in effective rail pressure and actual flow rate. This is covered later when we discuss forced induction.

Much like any other spring-mass system being driven by an electric current, opening of the injector is not an exact step change. It takes a small amount of time to build up enough energy in the coil to begin to move the pintle of the injector off the seat and allow fuel to flow into the manifold. The initial delay is known as "dead-time," but is also followed by a period of exponential movement of the pintle until it hits the fully open position. (**Figure 5-1**) Likewise, closing the valve takes time as well. Once the current to the coil is removed, the pintle is pushed back to its seat by internal spring pressure as well as fuel pressure behind it. The difference between the opening delay and closing delays is called "injector offset."

Remembering that fuel injectors are actuated by electromagnets, it is important to further understand how their performance can change. The strength of the electromagnet in the injector varies relative to voltage. (**Figure 5-2**) Having more voltage across the field of the coil increases the strength and allows the injector to open quicker. This in turn means that fuel begins to flow into the manifold slightly sooner if voltage is higher. Knowing that cars almost never have constant voltage, the PCM needs to be able to adjust. A failed alternator, dead battery, or even normal cranking can send voltage to 11.5 or lower. Normal charging usually keeps voltage around 14 volts, and a failed voltage regulator can send output above 17 volts. The bottom line here is that the same injector under these varying conditions can change its actual output by 40% or more. All modern PCMs have tables built into their software code to model this change, even if they aren't overtly visible to the calibrator. Various injectors exhibit different voltage compensation curves. While all injectors change relative to voltage in a similar manner, the exact offsets at a given voltage are slightly different as internal construction of the injector changes. To best model the actual fuel delivery to the engine, it is ideal to accurately input the voltage compensation for the injector used. A quick Internet search can often yield the exact voltage offsets for most injectors.

To add more complexity, the actual flow rate changes based on pulsewidth. As the injector first opens, more fuel flows for the split-second that pressure differences are the highest. (**Figure 5-3**) Additionally, when the PCM commands an injector-opening event for a short duration, there is a tendency for the injector's spring-mass system to overshoot the desired duration. The net result is that at small pulsewidths, the injector tends to deliver fuel at a rate slightly higher than the static flow rate of the injector. This often leads to modeling the injector with two different flow rates, one for the normal pulsewidths of cruising and power delivery and another for the shorter pulsewidths of idle and starting. This in turn leads to the need to determine where this change, known as the "break point," occurs. This break point is usually relatively

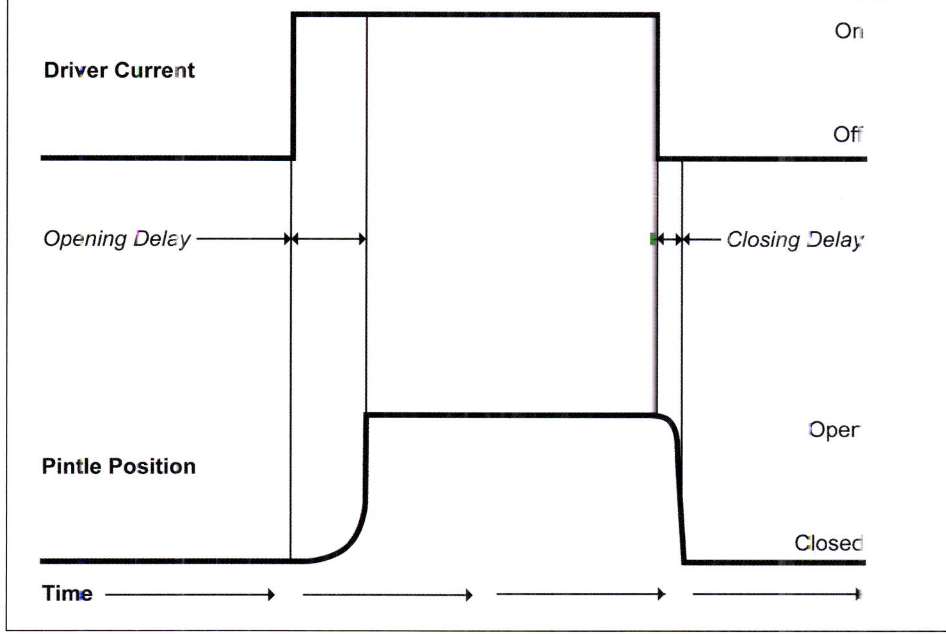

Figure 5-1 *A small delay occurs between when injectors are energized and when fuel actually begins to flow. Likewise, there is a similar lag between shutoff and closing. (Nate Tovey)*

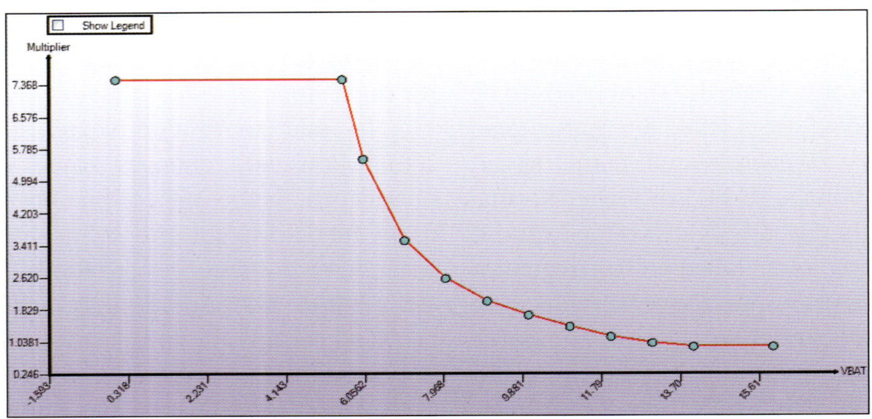

Figure 5-2 *Lower system voltages require additional time to open the fuel injector. This curve is unique to the injector design and should be adjusted when changing injectors.*

small, on the order of 1 to 3 ms, so the effect is often only seen at idle and very low loads. Again, entering this break point into the PCM routine allows for more accurate modeling of exactly how much fuel can be expected to enter the engine for a given commanded pulsewidth.

Most PCMs model the short pulse phenomenon with an injection time adjustment to the static flow rate model. In this case, the injector is treated as if the actual flow never changes. Instead, the commanded pulsewidth at low duty cycles is adjusted only by a time constant that becomes less dominant as injection time increases. In this case, a small amount of injection time (usually less than one ms) is added directly to the output under all operating conditions. The total fuel mass delivery effect of injector dead time is more pronounced at short pulsewidths, so many PCMs use this as the only means of compensation for short pulse fueling errors. This dead time characteristic is often easier to find than separate high and low injector slopes. It is important to recognize which method is being used when changing the injector characteristics, but either works fine.

Once all aspects of the fuel injectors' performance have been modeled in the PCM, it becomes much easier to actually deliver a precise amount of fuel to attain an exact A/F ratio. A poorly modeled injector can lead to sloppy A/F ratio control just the same as a faulty sensor. Although not all EFI systems have all of the fuel injector parameters visible to the tuner, most of them at least have some default values that get the calculations close enough to operate. In an ideal world, the injector parameter in the PCM is adjusted to exactly model the performance of the injectors being used. If this is done correctly, it should be possible to change only the fuel injectors and their associated parameters in the software on a perfectly tuned engine and see no change whatsoever in engine performance.

Ignition

Standard electronic ignition systems rely upon a switching circuit mounted inside the distributor itself to cycle power to the coil. Actual ignition advance is controlled by changing the phasing of the distributor relative to the crankshaft. Simply

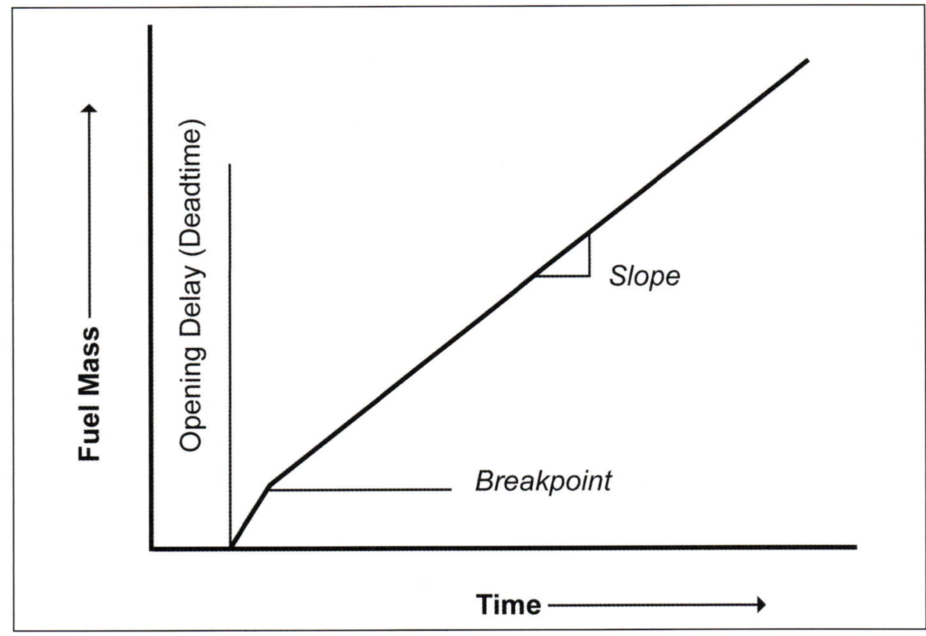

Figure 5-3 *Even though injectors are labeled with a flow rate "XX pounds per hour," we can see that this is only true for large durations. For short pulsewidths, some error must be calculated to adjust for deadtime and non-linearity. (Nate Tovey)*

rotating the distributor changes the delivered spark advance. Additionally, rotating sprung weights inside the distributor can change this phasing relative to engine speed to add timing at high speeds.

Modern EFI systems add another link into this chain. The PCM intercepts the ignition trigger signal from either a sensor inside the distributor or the crankshaft position sensor. It also has the capability to shift the phase of this signal (change actual ignition timing) before it is passed on to the coil. Since the PCM has the benefit of knowing exactly where TDC for cylinder number one is (thanks to the crankshaft position sensor), very precise control of the ignition timing

Removing the distributor from the chain of ignition components allows more precise control of spark events. These coil packs can recharge quickly enough to fire on both power and exhaust strokes, improving emissions. (Nate Tovey)

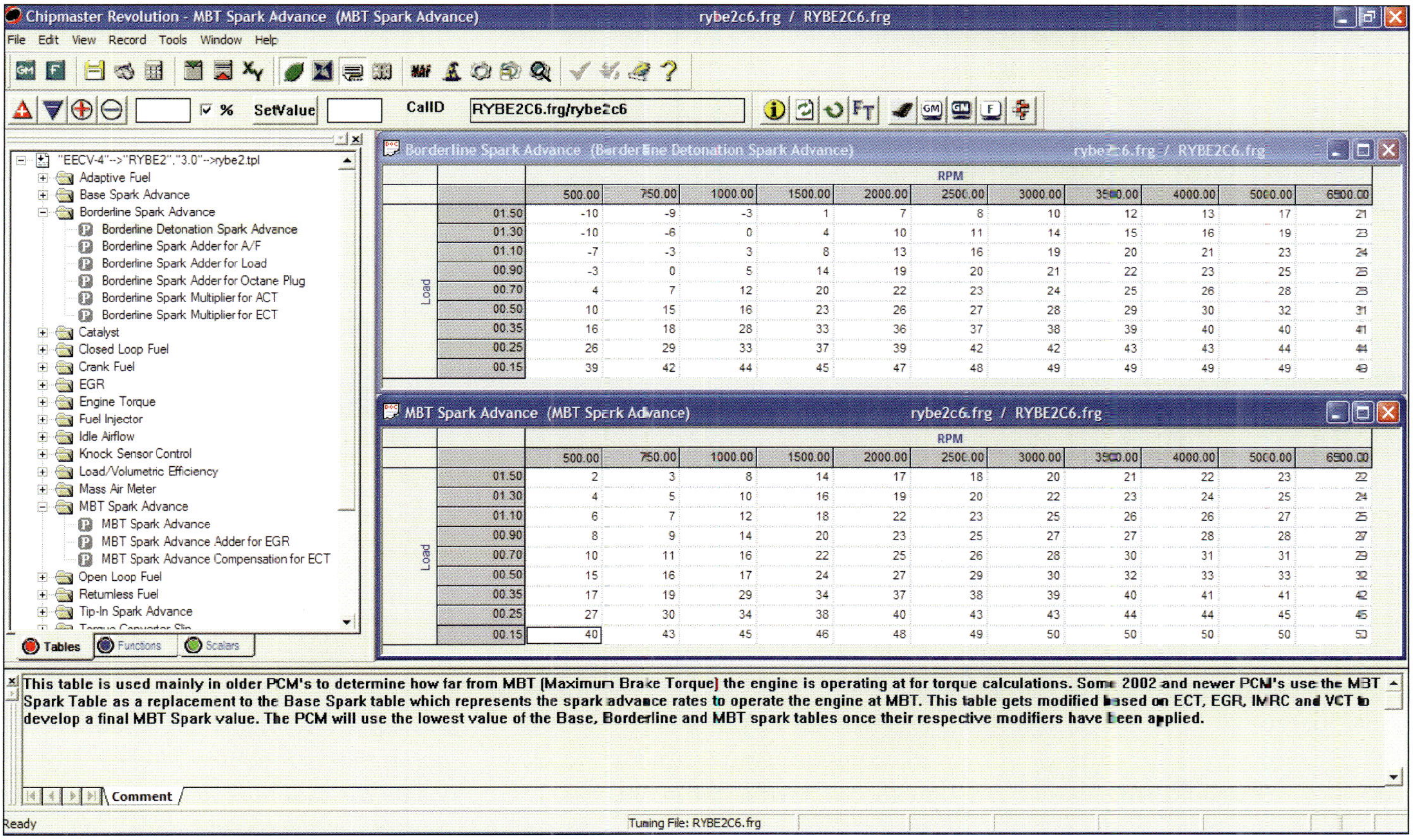

Chipmaster Revolution - MBT Spark Advance (MBT Spark Advance) — rybe2c6.frg / RYBE2C6.frg

File Edit View Record Tools Window Help

CallID: RYBE2C6.frg/rybe2c6

"EECV-4"-->"RYBE2","3.0"-->rybe2.tpl
- Adaptive Fuel
- Base Spark Advance
- Borderline Spark Advance
  - Borderline Detonation Spark Advance
  - Borderline Spark Adder for A/F
  - Borderline Spark Adder for Load
  - Borderline Spark Adder for Octane Plug
  - Borderline Spark Multiplier for ACT
  - Borderline Spark Multiplier for ECT
- Catalyst
- Closed Loop Fuel
- Crank Fuel
- EGR
- Engine Torque
- Fuel Injector
- Idle Airflow
- Knock Sensor Control
- Load/Volumetric Efficiency
- Mass Air Meter
- MBT Spark Advance
  - MBT Spark Advance
  - MBT Spark Advance Adder for EGR
  - MBT Spark Advance Compensation for ECT
- Open Loop Fuel
- Returnless Fuel
- Tip-In Spark Advance
- Torque Converter Slip

Tables | Functions | Scalars

Borderline Spark Advance (Borderline Detonation Spark Advance) — rybe2c6.frg / RYBE2C6.frg

Load \ RPM	500.00	750.00	1000.00	1500.00	2000.00	2500.00	3000.00	3500.00	4000.00	5000.00	6500.00
01.50	-10	-9	-3	1	7	8	10	12	13	17	21
01.30	-10	-6	0	4	10	11	14	15	16	19	23
01.10	-7	-3	3	8	13	16	19	20	21	23	24
00.90	-3	0	5	14	19	20	21	22	23	25	25
00.70	4	7	12	20	22	23	24	25	26	28	23
00.50	10	15	16	23	26	27	28	29	30	32	31
00.35	16	18	28	33	36	37	38	39	40	40	41
00.25	26	29	33	37	39	42	42	43	43	44	44
00.15	39	42	44	45	47	48	49	49	49	49	49

MBT Spark Advance (MBT Spark Advance) — rybe2c6.frg / RYBE2C6.frg

Load \ RPM	500.00	750.00	1000.00	1500.00	2000.00	2500.00	3000.00	3500.00	4000.00	5000.00	6500.00
01.50	2	3	8	14	17	18	20	21	22	23	22
01.30	4	5	10	16	19	20	22	23	24	25	24
01.10	6	7	12	18	22	23	25	26	26	27	25
00.90	8	9	14	20	23	25	27	27	28	28	27
00.70	10	11	16	22	25	26	28	30	31	31	29
00.50	15	16	17	24	27	29	30	32	33	33	32
00.35	17	19	29	34	37	38	39	40	41	41	42
00.25	27	30	34	38	40	43	43	44	44	45	45
00.15	40	43	45	46	48	49	50	50	50	50	50

This table is used mainly in older PCM's to determine how far from MBT (Maximum Brake Torque) the engine is operating at for torque calculations. Some 2002 and newer PCM's use the MBT Spark Table as a replacement to the Base Spark table which represents the spark advance rates to operate the engine at MBT. This table gets modified based on ECT, EGR, IMRC and VCT to develop a final MBT Spark value. The PCM will use the lowest value of the Base, Borderline and MBT spark tables once their respective modifiers have been applied.

Comment

Ready — Tuning File: RYBE2C6.frg

These MBT and Borderline knock timing maps show how many more speed and load points can be adjusted to have a unique timing setting than one could get with a weighted distributor. This table is shown using the Diablosport Chip-Master Revolution software.

events can be implemented. The PCM can then adjust ignition timing with respect to any other parameter it either has an input for or can calculate. Most PCMs have the main timing adjustments made based on a table of engine speed versus load, MAP, or airflow. Additional adjustments are usually made with respect to coolant temperature and air inlet temperature. When idling, small changes in ignition timing can be used to effectively control idle speed at a steady airflow rate. Dwell time of the ignition event can also be continuously adjusted in a similar manner.

The net output is still a switched ground to the ignition coil. Modern PCMs can also have paired outputs to drive a coil pack of two or more cylinders. Paired ignition outputs are used in what is called a "waste spark" ignition system. A waste spark system has a common coil for two cylinders, each opposite the other in firing order. These systems fire the spark plug on each cylinder once per revolution (twice per cycle). This yields a spark event during both the compression and exhaust strokes to help reduce emissions. A side benefit to this is that each coil doesn't have to fire all cylinders of the engine. This means slightly more time in between ignition events to allow the coil field to recharge to full energy. The final option is individual output for a separate ignition coil on each cylinder. These allow for more coil saturation time during high engine speeds and highest overall spark energy.

Fuel Pump

Most EFI systems are run with a "return fuel" style system where a pump in or near the fuel tank moves fuel to the rail on the intake manifold.

The latest form of complete PCM ignition control is seen here with the coil-on-plug system. Each spark plug is fed by its own coil with independent charging capability. (Nate Tovey)

Pressure is regulated at the rail and unused fuel is returned to the tank. It is important that the pressure regulator is set to deliver the same pressure with which the PCM is programmed to work. Changes in delivered pressure behind the injectors directly change their flow rate, making for inaccurate fuel delivery if the PCM does not compensate.

Some systems such as that found in GM's LS1 Camaros have a regulator and short return line near the tank with a single regulated line feeding the rail. This shorter return loop exposes less of the fuel to the heat of the engine compartment, reducing evaporative emissions concerns. These systems often have no connection between the intake manifold and fuel pressure regulator. In this case, the PCM assumes a constant rail pressure and calculates a differential based on

output from the MAP sensor. The PCM then adjusts pulsewidth accordingly to compensate for changes in manifold pressure.

"Returnless" fuel systems have only a single fuel line from the tank to the rail. In these systems, the fuel pump is not always operating at 100% output. The PCM controls the duty cycle of the pump based on the feedback from a pressure sensor mounted to the fuel rail. These pressure sensors are also usually attached to the intake plenum by a vacuum line to register fuel delta, or pressure drop, across the injectors. Since effective injector size varies with delta pressure, this allows the PCM to monitor conditions very closely. Additionally, the PCM can effectively make the injectors larger by commanding a higher delta pressure. Returnless systems usually have several tables to model fuel pump flow based on demand and available voltage. Fuel pump duty cycle in these systems may be as low as 10% at idle.

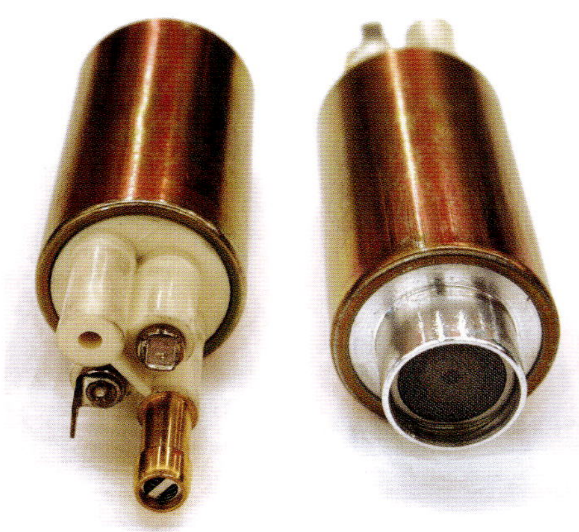

Modern EFI systems use a fuel pump that is mounted directly inside the fuel tank itself. This unit features a positive displacement gerotor design that is only compatible with return fuel systems that run the pump at 100% all the time. (Nate Tovey)

The fuel pressure regulator is usually mounted directly on the rail with a reference line to the intake manifold. As manifold pressure changes, the job of the regulator is to maintain the same pressure drop across the injector. This helps ensure that fuel is delivered at a predictable rate as the engine demands change. (Nate Tovey)

Throttle/ETC/Fly by Wire

The majority of vehicles on the road today have a cable connecting the accelerator pedal and the throttle blade. In this simple application, driver input directly equals actual throttle position. The TPS monitors this actual position and allows the PCM to adjust accordingly.

Newer cars are increasingly adding Electronic Throttle Control (ETC), or "fly by wire" features to the PCM. This system includes a Pedal Position Sensor (PPS) mounted directly on a spring-loaded accelerator pedal with no mechanical connection to the engine. The output of the PPS indicates the driver's intent to the PCM, where an actual throttle position is calculated and commanded to a stepper motor attached to the throttle blade itself. A TPS sensor is still used to monitor actual throttle position and allow for corrections by the PCM. This system allows the PCM to make adjustments to the throttle position to improve idle quality, off idle response, and drivability without the driver's help.

With the physical connection removed between the accelerator pedal and throttle blade, some redundancy and safety checks are now required. The most obvious is the addition of a second TPS sensor. The additional sensor uses either a different range or slope than the primary TPS and is used to verify actual position as well as perform a rationality check on the primary sensor output. Any errors between the two TPS sensors usually lead to a fault code and possible trigger of a failsafe mode.

The ETC strategy must also have some form of failsafe detection for stuck open throttle or improper operation at part load. Since there are no longer external springs mechanically forcing the cable to retract when the driver lifts off of the accelerator pedal, great care must be taken to make sure that the ETC's stepper motor is under control at all times. To prevent engine damage and reduce chances of a vehicle crash, this failing condition

ETC-equipped vehicles do away with throttle cables. The servo motor selects actual blade position, and can be seen here (arrow).

triggers a "limp mode" for the ETC system where the throttle position is limited to a very small value (usually less than 20%) until reset. ETC systems usually have a table in the ECM of maximum airflow rates at a commanded throttle position and engine speed. (See **Figure 5-4**) If the airflow exceeds the table's value, it is determined that something must be wrong with the ETC hardware.

When parts are changed on an ETC equipped vehicle in the name of adding horsepower, we run the risk of exceeding the PCM's predicted maximum airflow. It's not unusual to see an ETC equipped car suddenly drop into limp mode on the first test drive after adding a supercharger or larger camshaft. This is due to the added airflow. The solution is to reprogram the PCM to allow for higher airflow rates at high throttle before triggering limp mode. A safety margin of roughly 20% should be sufficient, but each vehicle can be unique. Only actual testing on the specific combination of vehicle and performance parts dictates the proper values to maintain safety with-

out inadvertently triggering limp mode under power. Again, great care must be taken not to simply disable this function. This would leave the potential failure of a stuck throttle and vehicle crash as the responsibility of the person who modified the PCM. If the OEM doesn't want to be taken to court, it's doubtful a smaller shop or individual tuner would fare well either.

Two main strategies for ETC control are used today: "pedal follower" and "torque-based." Pedal follower was the first ETC strategy to be widely employed and is far simpler in its operation. This throttle control strategy takes a PPS input from the driver and gives a reaction that is roughly equal to that at the throttle blade. The main exceptions are usually closed position for idle control and tip in rate to control driveline lash or "clunk." At idle, the ETC motor makes small corrections to the throttle blade angle to adjust effective throttle area. This change in area performs the function of a traditional IAC valve without the cost or complexity of another piece of

hardware. The PCM has the authority to make very fine adjustments that can effectively regulate airflow even at low engine speeds through the throttle blade alone. During upshifts, the throttle position can be reduced to briefly decrease engine power. This reduced output smoothes the transition into the next gear as the clutch engages, reducing driveline shock and noise.

With torque-based ETC control, the strategy is significantly more complex. Actual throttle blade position is no longer directly tied to the PPS. Instead, the PPS is interpreted by the PCM as a torque demand. The PCM has the latitude to choose how it delivers the desired torque. Actual engine torque can be manipulated by throttle position, spark advance, lambda, cam position, or any combination of these. The PCM must be programmed with multiple tables representing the engine's output capability under all operating conditions. During operation, the PCM is constantly calculating the exact engine torque output. Driver input changes are processed as torque demands the

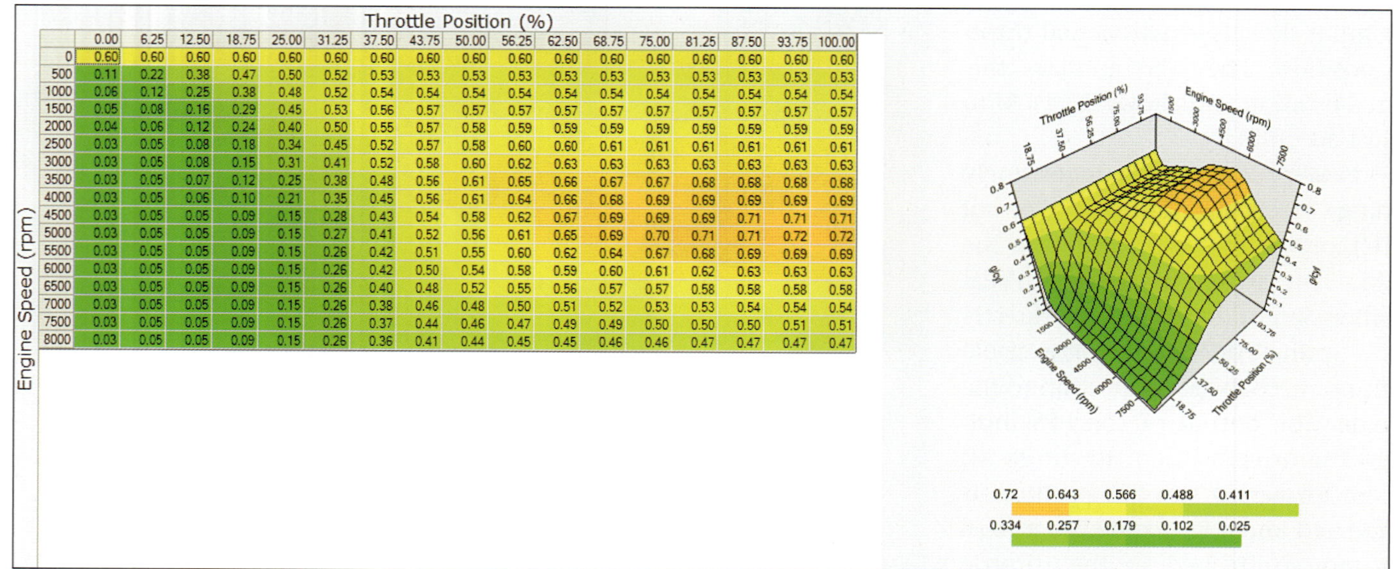

Engine Speed (rpm)	Throttle Position (%)																
	0.00	6.25	12.50	18.75	25.00	31.25	37.50	43.75	50.00	56.25	62.50	68.75	75.00	81.25	87.50	93.75	100.00
0	0.60	0.60	0.60	0.60	0.60	0.60	0.60	0.60	0.60	0.60	0.60	0.60	0.60	0.60	0.60	0.60	0.60
500	0.11	0.22	0.38	0.47	0.50	0.52	0.53	0.53	0.53	0.53	0.53	0.53	0.53	0.53	0.53	0.53	0.53
1000	0.06	0.12	0.25	0.38	0.48	0.52	0.54	0.54	0.54	0.54	0.54	0.54	0.54	0.54	0.54	0.54	0.54
1500	0.05	0.08	0.16	0.29	0.45	0.53	0.56	0.57	0.57	0.57	0.57	0.57	0.57	0.57	0.57	0.57	0.57
2000	0.04	0.06	0.12	0.24	0.40	0.50	0.55	0.57	0.58	0.59	0.59	0.59	0.59	0.59	0.59	0.59	0.59
2500	0.03	0.05	0.08	0.18	0.34	0.45	0.52	0.57	0.58	0.60	0.60	0.61	0.61	0.61	0.61	0.61	0.61
3000	0.03	0.05	0.08	0.15	0.31	0.41	0.52	0.58	0.60	0.62	0.63	0.63	0.63	0.63	0.63	0.63	0.63
3500	0.03	0.05	0.07	0.12	0.25	0.38	0.48	0.56	0.61	0.65	0.66	0.67	0.67	0.68	0.68	0.68	0.68
4000	0.03	0.05	0.06	0.10	0.21	0.35	0.45	0.56	0.61	0.64	0.66	0.68	0.69	0.69	0.69	0.69	0.69
4500	0.03	0.05	0.05	0.09	0.15	0.28	0.43	0.54	0.58	0.62	0.67	0.69	0.69	0.69	0.71	0.71	0.71
5000	0.03	0.05	0.05	0.09	0.15	0.27	0.41	0.52	0.56	0.61	0.65	0.69	0.70	0.71	0.71	0.72	0.72
5500	0.03	0.05	0.05	0.09	0.15	0.26	0.42	0.51	0.55	0.60	0.62	0.64	0.67	0.68	0.69	0.69	0.69
6000	0.03	0.05	0.05	0.09	0.15	0.26	0.42	0.50	0.54	0.58	0.59	0.60	0.61	0.62	0.63	0.63	0.63
6500	0.03	0.05	0.05	0.09	0.15	0.26	0.40	0.48	0.52	0.55	0.56	0.57	0.57	0.58	0.58	0.58	0.58
7000	0.03	0.05	0.05	0.09	0.15	0.26	0.38	0.46	0.48	0.50	0.51	0.52	0.53	0.53	0.54	0.54	0.54
7500	0.03	0.05	0.05	0.09	0.15	0.26	0.37	0.44	0.46	0.47	0.49	0.49	0.50	0.50	0.50	0.51	0.51
8000	0.03	0.05	0.05	0.09	0.15	0.26	0.36	0.41	0.44	0.45	0.45	0.46	0.46	0.47	0.47	0.47	0.47

0.72	0.643	0.566	0.488	0.411
0.334	0.257	0.179	0.102	0.025

Figure 5-4 *ETC systems employ a safety check table that defines the predicted maximum airflow at each speed and throttle point. If the actual airflow exceeds this table by too much, a fault is detected and the PCM reverts to failsafe mode.*

With the side cover removed, the drive mechanism of this ETC actuator can be clearly seen. Dual TPS sensors are integrated as well.

MAP sensor is then returned to the ETC strategy again to verify that actual engine load closely matches predicted load to catch any error conditions that may result in "undesirable" engine acceleration.

Modifying the engine, exhaust, or accessories to increase power directly changes the relationship between engine power and throttle position. Torque-based systems often recognize this extra power as a fault and trigger limp mode if the change is large enough. Even though the PCM has been reprogrammed to maintain the proper air/fuel ratio and spark to operate correctly, the ETC system may not allow for so much airflow. In some cases, as little as 5% more power may be enough to force limp mode. The solution is to once again look into the tables controlling ETC function and adjust them to reflect the power the engine is actually making. In some instances, it may be possible to change the ETC strategy from torque-based to

same way accessory inputs are. This makes for easier programming during the development phase for OEM engineers. The PCM can be programmed such that a certain A/C compressor requires 15 ft-lbs of torque to turn once engaged rather than fiddling with IAC or spark individually to smooth engagement.

To work correctly, torque-based ETC must start with an accurate representation of engine output. This is accomplished with a table representing actual engine torque (a derivative of airflow, lambda, and ignition timing) versus load and engine speed. To find the predicted load (air mass flow), another table is required to model airflow versus effective throttle area. The physics of this table are unique to the engine configuration. Changes in throttle body area (diameter), camshaft design, intake manifold design, or even exhaust backpressure can require changes to this table to prevent unwanted fault triggering.

The majority of engine calibration must be done prior to the final torque-based ETC calibration. This is accomplished by manually setting the ETC to a fixed position or range

and filling in the reference tables with actual values for steady state load versus engine speed and engine torque versus load and speed. Once these reference tables have been calibrated, a map can be built to translate a driver request from the PPS into desired engine output. Actual output as measured by the MAF or

Another ETC actuator (arrow) can be seen on this boxer engine. This one is tightly packaged between the intercooler and red intake manifold.

pedal follower to more easily accept modifications. Great care must still be taken to retain some form of genuine fault detection and control.

Since many transmission controls are also based on engine torque input, shift strategy is also optimized. Automatic transmission equipped vehicles with ETC usually take the torque control one step further to model instantaneous wheel torque output rather than just engine torque. Driver demands from PPS are calculated as wheel torque requests (vehicle speed or acceleration rate). With a known driveline ratio as indicated by the transmission gear selector and torque converter status, engine output torque can be recalculated as delivered wheel torque. This allows OEM calibrators the luxury of adjusting both engine and transmission conditions to provide a combination that supplies the desired wheel torque the driver has requested while performing earlier converter lockup and upshifts at light load to improve fuel economy. If a driver's demanded torque exceeds the available reserve in the current gear at WOT, a downshift can then be performed to acquire a higher ratio. Another side benefit is the possibility of executing a transmission shift with zero change to wheel torque by manipulating engine output before, during, and after the shift. This zero torque change shift can yield a seamless driving experience to even the most discriminating consumers, a big benefit for OEMs looking for minimal NVH.

Idle Air Control

The throttle blade is the largest controlling factor in an engine's airflow, but at low flow rates such as idle it may not be practical to accurately

torque-based ETC control is not a roadblock to making big horsepower. This 2005 Mustang is supercharged and retains stock drivability with the torque-based ETC system, even though it makes over 200 hp more than a stock configuration. (Nate Tovey)

control the airflow rate with one large opening. The idle air control (IAC) valve acts as a small throttle valve with a narrower range of operation but finer control. IAC valves are multi-position devices usually based on a stepper motor that moves a plunger to cover a portion of an alternate flow path around the throttle blade. The PCM has a table showing the transfer function of the IAC as command position versus actual bypass flow. The PCM has a variable output that it uses to command a flow rate through this valve to control idle speed, idle stability, and deceleration. The IAC acts as

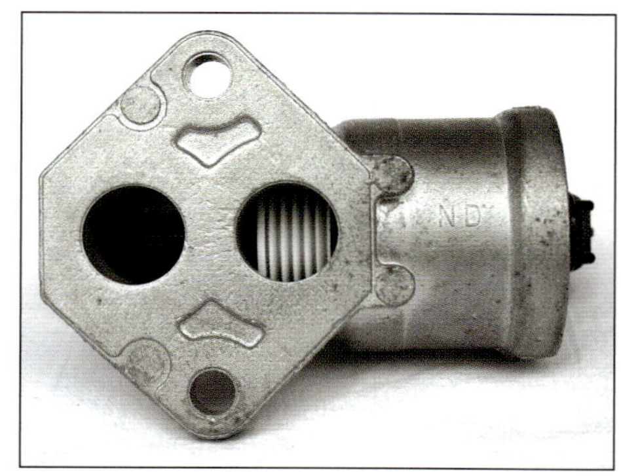

The IAC motor acts as a small throttle body. By changing its position, the amount of bypass air from one side to the other is incremented. (Nate Tovey)

The IAC motor can be seen here as installed on the side of the throttle body assembly. Air can be diverted directly from the front to the rear side of the throttle blade. (Nate Tovey)

The MRT "Interceptor" was one of the first supercharged ETC-equipped Mustangs to be displayed at the SEMA show. (Nate Tovey)

the "coarse" tuning control of idle speed and stability. Since the valve cannot instantaneously open or close, it is used to control predicted steady state airflow during idle. Faster response to changing idle speed is done by spark advance since this can be done as quickly as the next ignition cycle while the IAC is still moving. During deceleration, the IAC acts as the dashpot control (more on dashpot in Chapter 10). If the PCM commands a lower output for IAC during closed throttle, the engine loses speed more quickly. This can also be adjusted to change the amount of engine braking being done during coast down.

Runner Controls

Some engines are equipped with intakes that include multiple possible flow paths for incoming air. Longer runners are used to generate better torque at low and mid RPMs. Shorter runners are used to increase the ram tuning effect at high speed. Changing between the two flow paths is done by opening a butterfly valve attached to a solenoid. The point at which the sole-noid opens the runner control valve can be dictated by the PCM as a function of any other input, usually TPS and RPM.

Another more recent innovation has been the implementation of charge motion plates or "port flaps." These are similar in construction to runner controls, except that they only restrict flow across a single runner rather than direct flow from a different runner altogether. The purpose of a charge motion plate is to direct the intake charge toward one side of the intake port, improving the amount of mixing that happens once the air enters the cylinder itself. These plates are usually closed at idle and low speeds, leaving a small opening that results in a higher velocity stream of air heading toward the intake valves. Once engine speed (and subsequently, port velocity) picks up, the plates are opened similar to a throttle blade to allow virtually unrestricted flow. Generally speaking, very little extra power is gained by removing these from the intake tract, and the low-end torque loss is usually significant enough to be noticed by the average driver.

Cam Controls

Camshaft controls are quickly becoming commonplace in modern engine controls. The ability to change valve-timing events on the fly allows automakers to tune engines more optimally for horsepower, emissions, and fuel economy simultaneously. These systems work by placing an oil-fed valve and spring assembly on the camshaft drive mechanism. A remote PCM-triggered solenoid directs oil flow to the assembly overcoming spring tension and changing actual valve timing.

Honda's VTEC system was an early pioneer where a change from one set of cam lobes to another made for a distinct step change in engine output. Essentially, operating on a small camshaft at idle, cruise, and lower RPMs made for good mileage and emissions, and changing to larger cam lobes as speeds increased made more power.

BMW's VANOS system was a leader in the more widely used cam phasing approach. This system uses a single set of cam lobes where the indexing of the cam relative to the crankshaft can be adjusted. By shifting the cam phasing relative to the crankshaft, it becomes possible to keep the ram tuning effect more powerful across a broader RPM range, increasing overall efficiency. Cam phasing has the benefit of being linear as opposed to the older binary VTEC system. With finer control and more adjustment points, camshaft performance can be optimized rather simply by switching between two set points.

On DOHC engines, it becomes possible to rotate intake and exhaust camshafts independently of one another to allow for another degree of freedom. With dual cam phasing, not only can each cam be indexed relative to the crank, but now overlap can also be incremented. With independent control, the exhaust duration can be adjusted at cruise for increased EGR or the intake cam can be adjusted to increase cylinder pressure as speed and loads change. Overlap can be added at high RPM to increase ram-tuning magnitude without creating unnecessary EGR flow or loss of compression at idle where the lobe centers can be separated more. Some variable cam timing systems also increase overlap during extreme cold start conditions to reduce pumping effort and use the warm exhaust to impart heat to the frozen intake manifold. Once a predetermined temperature is reached, the system reverts to a wide lobe separation to stabilize idle and reduce the amount of intake charge blowing from the intake to the exhaust without combustion.

BMW has further upped the ante lately with the addition of complete lift control. By adding another solenoid on a cam that changes the valve lift with respect to camshaft lobe lift, anywhere from 0 to 100% of the camshaft's lift can be applied to the valves themselves. This has the potential to eliminate the need for a throttle blade since the intake valves themselves become the throttling mechanism to limit engine airflow. Although not currently in use for high-performance vehicles, the technology shows incredible potential that even the horsepower junkie can appreciate.

Boost Control

On supercharged and turbocharged engines the throttle blade is not the only method of load control. The amount of positive manifold pressure (boost) that an engine makes is directly proportional to the total engine load at a given operating point. Most OEM vehicles with forced induction include some form of PCM controlled boost limit. On a supercharged engine this is usually a bypass valve that can recirculate air around the compressor. Turbocharged engines typically employ a PCM-controlled wastegate to dump excess exhaust energy, limiting shaft speed. Either way, a PCM-triggered solenoid usually resides in series with the vacuum line between the reference (intake manifold or compressor outlet) and control (wastegate or bypass diaphragm). Most OEM PCMs employ some form of boost control to soften torque onset for driveline durability and NVH, manage "abuse," or provide for a limp mode without boost. These situations can prove detrimental to making more power than stock and usually require changes to the PCM to overcome.

This solenoid (arrow) activates a bypass control valve that can be used to reduce the manifold pressure in a supercharged engine to limit torque or heat. (Nate Tovey)

Camshaft phasing relative to the crank is accomplished by activating a solenoid to divert oil flow into the phasor. Oil is fed to the actuator from a solenoid immediately behind the cam gear. (Nate Tovey)

The EGR valve on this Subaru is installed on the back of the intake manifold. Hot exhaust gases are fed to the valve through the shielded steel pipe below the valve. (Nate Tovey)

"THE RECIPE"

Now that we have covered how information travels between the engine and PCM, it's time to start processing. Internal combustion gasoline engines only operate within a specific window. Things have to happen in the right order with the right amounts of each input for the desired result. Much like baking a cake, there is a recipe for turning gasoline into usable power. Whether you are making yellow sheet cake or triple chocolate surprise depends upon the engine and parts with which you start. Calibration cannot miraculously add airflow. That is up to the mechanicals in which the limit resides. However, just like forgetting to stir the mix properly or setting the oven temperature too high ruins the baked goods, the calibrator has the ability to prevent an engine from running well. Likewise, a good cook can bake the moistest cake with the best frosting design, the calibrator can polish the engine's performance to perfection as well.

All internal combustion, spark-ignited engines have a few rules that must be obeyed.

Combustion only occurs between 8.0:1 and 25.5:1 air/fuel ratios. The closer to the edge of this envelope the engine gets, the more likely misfire becomes. The stoichiometric mixture is 14.68:1 ($\lambda = 1$), which would theoretically leave the least amount of leftover ingredients. Slightly leaner mixes result in better fuel economy and slightly richer mixes result in better torque. Excess fuel can be used as a cooling agent to limit combustion temperatures at high load.

Peak flame speed is found at $\lambda \approx 0.9$. At this point, less ignition lead is required to reach peak cylinder pressure at the same time in the cycle. Adding excess fuel or air slows down the combustion process. When operating under significant power enrichment ($\lambda << 0.9$), leaning the mix out increases flame speed as we approach $\lambda \approx 0.9$ again. Additional fuel enrichment (to ratios below $\lambda \approx 0.9$) slows

Running an engine properly is much like making cookies. If the proper amounts of all components are delivered and mixed correctly, the results are more impressive. (Nate Tovey)

the flame front's travel, requiring more ignition advance to achieve peak cylinder pressure at the correct time.

The spark initiates the combustion process in normal operation. For this to happen, the spark energy must be high enough to create an arc across the plug gap. As the air/fuel charge mixture density increases, more energy is required to create the electrical arc in the gap. Charge density is directly proportional to engine load, so it is affected by throttle position, port velocity, compression, and manifold "boost." The wider the gap, the more energy is required to jump it. More surface area of the arc from a wider gap yields better initial combustion due to an increased number of oxygen and gasoline molecules in contact with the arc. Ignition energy is directly propor-

The importance of efficient manifold tuning can be seen in this BMW design. Long individual runners are sized to improve midrange torque of this engine, and the plenum can clearly be seen as well. (Nate Tovey)

tional to input voltage, coil winding, and coil saturation time.

Engine speed is a result of the difference between loads (friction, accessories, pumping losses, vehicle movement) and output (current engine power, often a result of throttle position). Any time output exceeds the current loads, speed increases. When output is less than current loads, engine speed drops. Holding a constant engine speed requires making just enough power to equal the current loads on the engine.

Idle speed is a perfect example of this delicate balance since load changes from accessories and friction as a function of speed must be offset by careful adjustment of engine power from throttling and spark. Low speed idling is one of the most difficult balances for any engine to make. EFI systems can experience significant trouble due to the relatively large amount of time between power strokes when attempting to make relatively small corrections. If the loads on the engine exceed output by enough, stalling occurs.

Spark advance at most operating conditions should be set as close to MBT (maximum brake torque) as possible to take advantage of maximum engine efficiency. It is difficult to damage an engine from over-spark at light load. MBT may make best power at WOT, but it is often wise to retard timing slightly to provide some safety margin and avoid knock. It is not ideal to operate at MBT at idle. Timing should be intentionally slightly retarded from best torque to allow for spark trimming of idle speed. If an engine is idling with a spark advance equal to MBT, it is not possible to add timing to quickly compensate for any added load or drop in speed.

Most engines are perfectly happy to run at stoichiometric (λ = 1, 14.68:1 A/F) ratio 95% of the time. The primary exception is WOT performance, where a richer mixture not only makes more torque, but also allows for more spark lead and cooler exhaust temperatures. Once the fuel mixture is right, adjusting spark for best operation becomes much easier.

Some vehicles with excessively large duration camshafts simply do not idle well at λ = 1, and adding up to 10% fuel can help idle quality. This increase in fuel delivery yields a slight torque increase which helps stabilize combustion, even at light load. These same vehicles are typically fine at λ = 1 for cruise, but require larger amounts of acceleration enrichment (AE) to prevent stumbling on tip in.

The bottom line is that once the calibrator has given the engine the right set of operating parameters, vehicle performance should be a direct result of the parts of the system. Too large of a camshaft and the calibrator has little hope of good idle quality. Too small of an intake runner and total power suffers regardless of what air/fuel ratio or spark lead is run. Keep in mind as a calibrator that it is possible to adjust things to a certain extent in the name of driving behavior, but there are limits. Often it's best to recognize the situation for what it is and either change parts or deal with the less-than-optimal results.

Modeling Airflow

To begin the actual calibration process for an engine, one of the most important steps is to recognize the instantaneous airflow. This airflow can then be processed to determine the necessary fuel delivery to maintain smooth engine operation. The most important core function of any PCM is this airflow modeling, which determines subsequent fuel commands. To this end, the calibrator must create adequate representations in the PCM code of what is happening in the engine's physical environment. Modeling of the engine airflow is done in one of two methods: Mass Air Flow Measurement or Speed Density Calculation. Either system is capable of properly controlling an engine, has its own pros and cons, and sometimes both are used together.

Mass Air Flow

Mass Air Flow systems rely heavily upon input from the MAF sensor discussed earlier. These systems take the output of the MAF sensor as a direct representation of current engine airflow. This approach makes for very simple and straightforward calculation of engine load and fuel requirements. In this case, engine load (Volumetric Efficiency) can be instantly shown as:

$$VE = \frac{MAF}{(Displacement \times \rho_{STP} \times Speed)}$$

Changes in throttle position simply restrict airflow. Part load (vacuum) is seen by the PCM simply as a smaller mass flow value. Since forced induction engines end up moving more total air mass per cycle, actual boost pressure is not required to calculate load and fuel demands. Knowing exact air mass flow makes fuel demand calculations simple as well:

$$Fuel\ flow\ rate = MAF \times (desired\ A/F\ ratio)$$

The crucial point to making mass air systems work properly is that the output of the MAF sensor must reflect reality. This is where the calibrator must spend some time to ensure correlation between actual mass flow and indicated mass flow. A large portion of the calibration process on a mass air system is spent tweaking the MAF transfer function in the PCM to reduce the variation across a wide range of steady state conditions. Making the MAF data more reliable to the PCM forces many other operating conditions to simply fall into place easier.

Any errors between the MAF output and actual engine airflow directly result in inaccuracy in fuel control and engine operation. The MAF must have adequate resolution, range, and repeatability for the system to work properly. This is particularly important when high RPM, high load flow rates have the potential to exceed the measurement range of a particular MAF sensor. In performance applications, it is not unusual to see the "pegged" MAF reporting a constant maximum value to the PCM. In turn the PCM commands a constant fuel delivery against what is actually an increasing airflow. The result is a progressively leaner air/fuel ratio and often detonation. The solution is either a change in MAF sensor hardware or some creative compensation in the PCM tables.

The placement of the MAF sensor in the inlet tract can create some challenges. As discussed earlier, the MAF should be positioned such that it sees laminar flow across its element. This often means that some tight inlet routes may not be conducive to proper MAF placement. Many aftermarket engine packages such as "spider EFI" intakes with a single throttle plate under a central air filter do not even leave any room for a MAF sensor. The only way to run a mass air system on these applications is to extend the inlet to a remote MAF and filter assembly, making for a somewhat more complex inlet system.

The primary benefit to the mass air system is that any changes in actual airflow that fall within the MAF sensor's range and resolution can be instantly accommodated by the PCM.

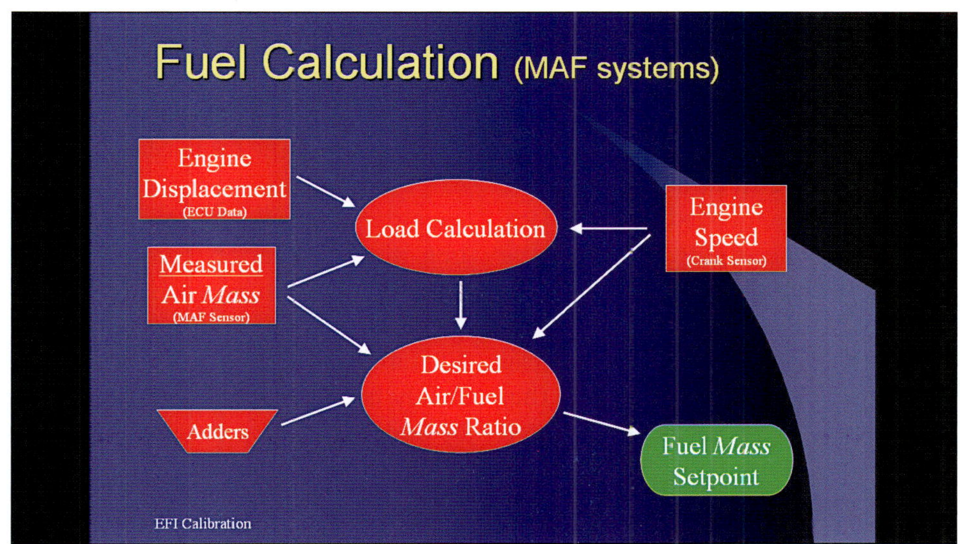

For a mass air based system, the amount of fuel delivered is calculated directly from the MAF sensor input and desired ratio.

This means that a change to a larger camshaft or higher flowing intake manifold simply show up to the PCM as slightly higher airflow rates, resulting in slightly higher fuel delivery. Mass air systems tend to be very forgiving of relatively drastic modifications in the name of horsepower. If it is possible to cleanly install a MAF sensor, this is the most desirable method of engine control due to its flexibility and accuracy.

Speed Density

The second strategy for engine control is speed density. In this method, airflow is never actually measured. Instead, incoming air mass is calculated based on temperature, manifold pressure, and engine speed, using a reference volumetric efficiency table. The engine's volumetric efficiency is a constantly moving value based upon RPM, camshaft design, manifold design, displacement, compression, and instantaneous manifold pressure. The main volumetric efficiency table is usually shown as manifold absolute pressure versus engine speed. The values contained in this table represent a volumetric fill percentage for the current manifold pressure and engine speed. The PCM must then take this volume, the MAP value, and inlet temperature to calculate the incoming air mass. High school physics has conveniently shown us the universal gas law: PV = nRT. Using this equation, the PCM can calculate air mass based on sensor inputs, molar density of air, and the gas constant for air. Once the incoming air-mass flow rate is calculated, a corresponding fuel flow rate can be determined. It is up to the calibrator to develop the volumetric efficiency table for each engine combination. Since there may be sim-

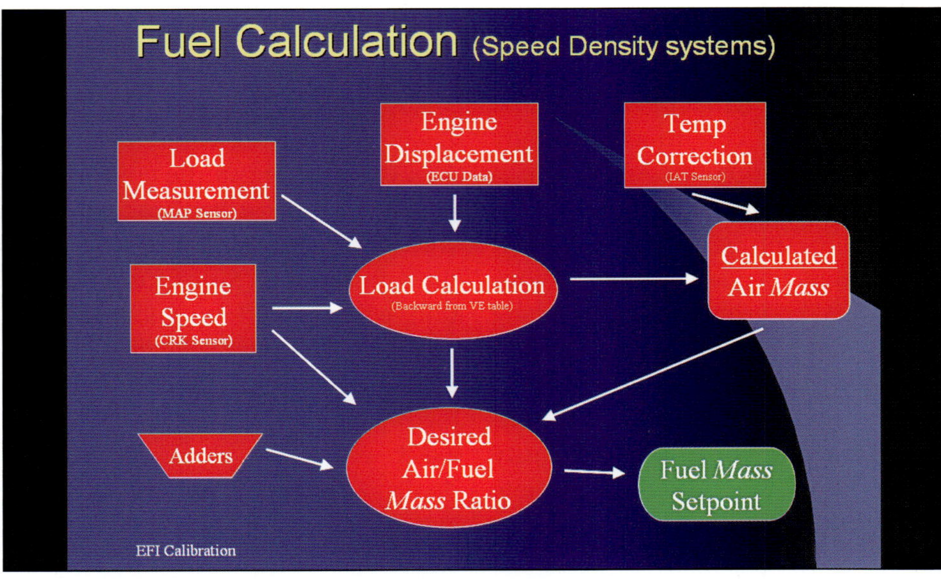

For a speed density system, air mass must first be calculated from a reference table before proceeding to the fuel calculation.

ilar airflow rates at various speed and MAP loading conditions, it is important to find steady state VE values for all regions of the table.

Temperature correction becomes critical in a speed density system. Unlike the mass air systems that have meters compensated for ambient changes, actual intake air density must be calculated in a speed density strategy. Changes in inlet temperature can have a significant effect on actual inlet air mass for a given volume and pressure. Boyle's gas law is hard at work here.

The downsides to speed density are a reduction in airflow resolution, increased time needed to calibrate, and reduced flexibility. Building the volumetric efficiency table typically takes longer than modeling the transfer function of a MAF sensor. Special care should be taken to ensure that the volumetric efficiency table is a relatively smooth function in both axes. This table is a representation of the engine's pumping characteristics and is generally smooth with respect to pressure and speed, just like any other

pump. Significant step changes to cam timing events, such as activation of VTEC, may dictate relatively large breaks in the RPM curves, but this is the exception.

Most volumetric efficiency tables are 32 x 32 or 16 x 16; many older versions are even smaller. At wide open throttle where MAP values are relatively constant, this leaves fewer adjustment points across the RPM range to model the engine's behavior. Although this function is relatively smooth, it still leaves quite a bit of interpolation to be done by the PCM between cells. The more cells available to model the engine's VE, the smoother overall operation can be, but the calibration process takes longer to optimize each additional column or row of cells. Once the base VE table has been constructed at a steady inlet temperature, testing must be done to develop the inlet temperature compensation curves. These temperature compensation curves may vary from the ideal thermodynamic curves due to sensor placement, construction, or reaction times.

One of the largest concerns to the performance calibrator is that speed density systems are based upon the assumption that engine volumetric efficiency remains essentially unchanged at each cell in the VE table. Changing engine components, especially camshafts, intake manifolds, or superchargers, can have a drastic effect on engine volumetric efficiency. Any change that significantly improves engine airflow requires another change to the base VE table to allow the engine to maintain the same relative fuel delivery. Since spark tables are also usually linked to engine load, speed density systems are especially prone to poor operation after parts changes without recalibration. A simple cam change may require several hours of dyno time to remap the VE table correctly. Additionally, if the range of the MAP sensor is exceeded by increasing manifold pressure (super or turbocharging), it becomes necessary to install a sensor with a wider range and rescale the pressure axis of the VE table to accommodate.

Speed density systems are popular because of their relative simplicity compared to mass air systems. By eliminating a fairly obtrusive sensor requirement in the inlet tract, designs become much more open. Exposed air boxes, filter elements, or individual stacks no longer pose a problem for air measurement. Many OEMs employ speed density because of the reduced cost of the eliminated MAF sensor. Although it may take slightly more development work to optimize the VE table, their economics of scale more than make up for it.

An even simpler version of speed density calculation used on some race-only vehicles is Alpha-N calculation. In this strategy, the base airflow map is a function of throttle position and engine speed. Without regard to actual manifold pressure, the calculation is simpler, but lacks resolution to correct many drivability issues or provide decent emissions control. This strategy is best left to dedicated drag cars where the engine's duty cycle is little more than idling in the pits and WOT down the track.

Fuel Delivery

The PCM only has an output for individual injector controls. Several steps must be performed to get from the desired air/fuel ratio to injector output. Once the PCM knows the incoming air mass, it is possible to calculate the desired fuel mass to attain the target air/fuel ratio. The first basic calculation is to determine desired fuel mass delivery:

$$\text{Fuel Mass (lb)} = \frac{[\text{Intake Mass Air Flow (lb/hr)}]}{[(\text{des. A/F ratio}) \times 120 \times (\text{RPM})]}$$

To calculate actual fuel delivery to the engine, one of the following formulas are used: see **Figure 6-1**

Once a desired fuel mass has been determined, a corresponding injector output can be calculated. The actual injector on time is adjusted based on static flow rate, rail pressure, and injector opening characteristics to convert desired fuel mass into a binary injector output for a fixed amount of time. Knowing the injector flow rate and coil characteristics is critical in this stage. Much like errors in the calculated intake airflow, errors in calculated fuel flow versus actual result in less than ideal operation and drivability. Whenever injectors are changed on a vehicle, it is important to update the injector variables in the PCM to reflect the all-new operating characteristics. This gives the PCM more accurate control over actual fuel delivery rather than simply assuming that a Bosch 24lbs/hr injector and Siemens 60lbs/hr injector behave identically with exception of their static flow rates. While the trend may be similar for both injectors versus voltage, the actual offset can be enough to cause a rough idle or poor transition behavior on tip in. Keep in mind that changes in fuel rail pressure or delta pressure across the injector have an effect on static flow rate, but not on the opening characteristics.

With accurately modeled airflow and fuel flow, calibration of transient conditions becomes much easier when the tuner no longer needs to chase his tail on small fuel calculation errors. The calibration process itself should be more focused on moving a desired and actual air/fuel ratio simultaneously rather than trying to make them converge on the fly. Worse yet, attempting to reach an actual air/fuel ratio by commanding a different desired ratio makes for a frustrating exercise when trying to calculate

Figure 6-1

$$\text{Mass of Fuel Delivered (lb)} = \frac{[\text{Injector Flow Rate (lb/hr)} \times \text{Pulsewidth (in milliseconds)}]}{(3.6e^6)}$$

OR

$$\text{Mass of Fuel Delivered (lb)} = \frac{[\text{Injector Flow Rate (lb/hr)} \times (\text{Duty Cycle}) \times 120]}{(\text{RPM})}$$

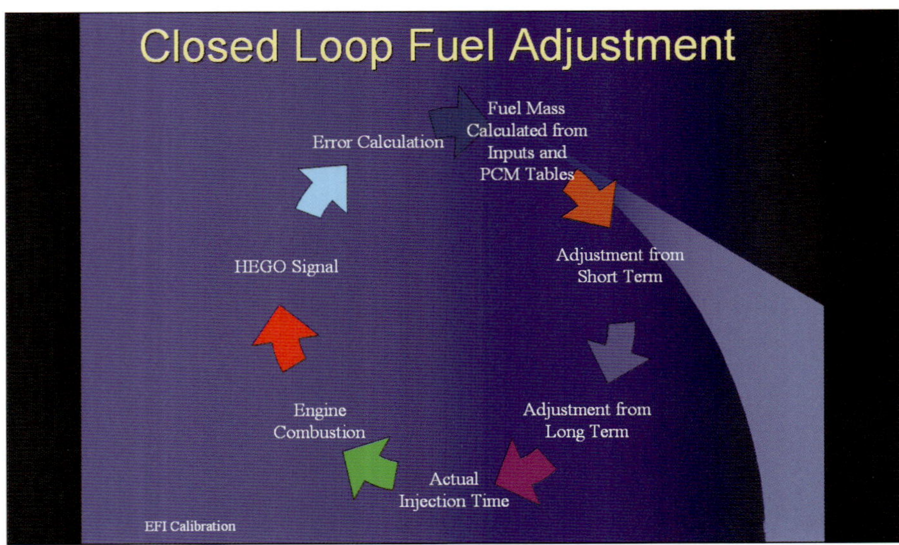

Closed loop systems constantly monitor engine gas outputs and adjust future fuel delivery accordingly. This allows the PCM to learn trends in engine performance, maintaining proper engine operation and emissions.

minor changes. Calibration is much easier when a 1% change in airflow comes with a 1% change in fuel flow.

The time at which the fuel is injected during the cycle can also have a pronounced effect on engine performance and emissions. Most OEM sequential EFI systems are timed such that fuel is injected against a closed intake valve slightly before opening. This allows the fuel to evaporate for better mixing and cools the valve simultaneously. The trick is to time this such that the evaporated fuel does not have time to travel back up the manifold runner before the valve opens. Many aftermarket "tuners" make the common mistake of trying to align injection timing directly with intake valve opening. Injecting fuel directly into an open valve can have the negative effects of poor mixing due to lack of evaporation. Low air and engine temperatures make the problem worse, as evaporation is slower at reduced temperatures. The unburned fuel and generally poor combustion yields a loss of engine torque along with excessive hydrocarbon emissions. After prolonged operation, improper injection timing may even lead to washing the oil coating off the cylinder bore and premature engine wear or poor ring sealing. The loss in torque alone from poor evaporation should be enough incentive to the performance calibrator to seek proper injection timing.

Picking a Ratio

Now that we have accurate control of the actual air/fuel ratio, choosing what that ratio will actually be for the engine is far easier. The next question becomes, "What air/fuel ratio is desirable?" The answer is, "That depends." Air/fuel ratio has a significant impact on emissions, fuel economy, idle quality, and power.

As mentioned earlier, stoichiometric mixture leaves the least amount of leftover primary reagents for the catalyst to process and generally gives good economy. Overall emissions are more involved due to the fact that different byproduct amounts vary with respect to air/fuel ratio. Hydrocarbon and CO emissions generally decrease with leaner mixes, with most significant reduction at $\lambda = 1$, and above. Because of the higher burn temperatures NOx emissions increase significantly above $\lambda = 1$. Although these leaner ratios can help fuel economy at cruise, modern emissions standards

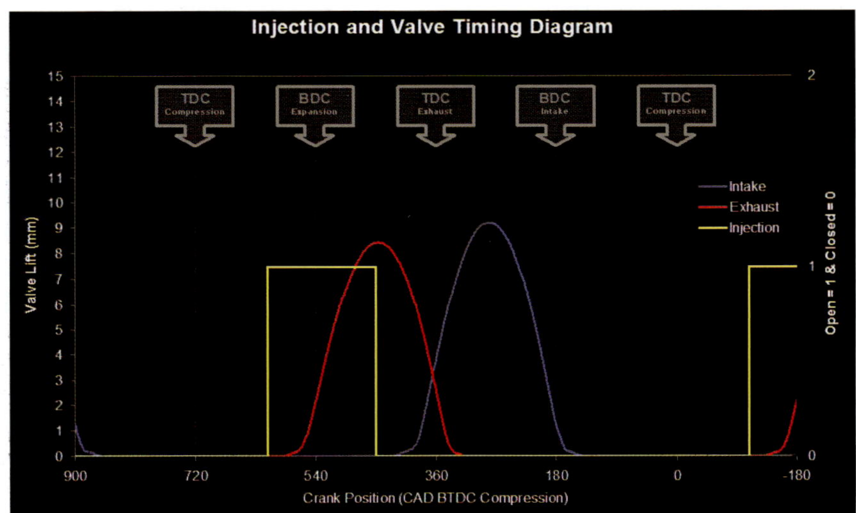

The timing of the injection can have a pronounced effect upon emissions and torque. This sample shows fuel being injected prior to intake valve opening in order to allow for complete evaporation before ignition.

often prevent continued operation above $\lambda = 1$.

As mentioned earlier, idle is often best done at $\lambda = 1$ for a number of reasons. Fuel economy and emissions are good, but there is still room for torque control (and consequently speed control) through changes in fueling from stoichiometric if necessary beyond the normal adjustments from the IAC or spark. In cases where camshaft design is too aggressive to support a stable idle at $\lambda = 1$, slight enrichment often stabilizes things. There is a practical limit to this enrichment from the potential for bore wash. If the idle mixture is too rich, fuel literally washes the walls of the cylinder clean of oil. Bore wash is most pronounced at idle due to relatively low intake port velocity combined with constant injector spray intensity. Without a large enough air charge to disperse and dilute the fuel spray, much of it lands on the opposing cylinder wall, rinsing off the oil layer. The result is excessive ring wear and eventual loss of compression due to blowby. Although the engine often idles smoothly as rich as $\lambda = 0.7$, anything richer than $\lambda = 0.85$ runs an increasing risk of bore wash, depending on port design. Conversely, an excessively lean condition at idle tends to exhibit a surging speed as torque changes when the leaner mix brings the engine closer to MBT. Although far less healthy for ring durability, the richer condition at idle is more stable and therefore helpful when attempting to initially calibrate idle speed without stalling. It is not recommended that the engine be allowed to operate at an excessively rich idle for more than a couple minutes without clearing any built up fuel deposits.

At wide-open throttle, the primary concern becomes torque output.

Remembering that a slightly rich mixture of $\lambda \approx 0.91$ (13.2 to 13.4:1 air/fuel) yields best torque, this becomes a good starting point for WOT air/fuel targets. Some OEMs continue to target $\lambda = 1$ at WOT to improve emissions, and are knowingly sacrificing potential power to do so. In many cases, knock limits or exhaust temperatures dictate that some additional fuel must be added at WOT. Richer mixtures burn cooler and consequently allow for more spark advance at the same load. These cooler burning rich mixtures also reduce the amount of heat to be absorbed by the cylinder head, piston, and valves. A cooler head and piston are less prone to preignition due to hot spots. Valve and catalyst durability is greatly reduced if temperature increases too much, so high mileage warranty concerns often drive OEM fueling requirements at WOT. For this reason, many OEM vehicles sold today approach $\lambda \approx 0.82$ (12.0:1 air/fuel) or richer once a catalyst protection threshold has been reached. Unlocking control of this protection circuit can be the key to properly calibrating such vehicles.

Adding extra fuel is a quick method of maintaining power output without reducing spark advance... to a limit. A little bit of extra fuel can go a long way toward providing a knock safety margin for the occasional bad tank of fuel. Many forced induction engines can simply not operate above $\lambda \approx 0.85$ without detonation due to the extra heat present in the intake charge as a result of the primary compression stage of the super/turbocharger. It has been my experience that most properly designed forced induction engines exhibit excellent power with good durability at approximately $\lambda \approx$ 0.8 (11.7 air/fuel ratio). This usually leaves enough cooling effect to allow for slight fuel and weather changes if calibrated at least 3 degrees away from the spark knock limit.

Going too far in this direction can also have a negative effect. Excessive over-fueling can result in rich misfire conditions or fuel burning off in the exhaust system, increasing EGTs. In extreme cases, fuel collects on the spark plug fouling the leads and hampering ignition performance.

Transients and Modifiers

To more accurately predict the dynamic fuel needs of an engine, EFI systems employ a model to predict the volume of fuel trapped on the manifold wall. This film is known as "Tau" (τ). Since no current port injection system has the ability to spray 100% of its fuel into the cylinder from a single injector pulse, part of this spray collects on the walls of the intake port. Incoming air flowing past this "puddle" of fuel on the port walls carries part of it away into the cylinder as well. The more volume of air flowing past the τ puddle, the more this puddle is being decreased in mass. Tau modeling systems use instantaneous airflow predictions to attempt to maintain a constant volume of τ. This means that increases in airflow require an increase in fuel to maintain constant τ volume and decreases in airflow allow for a decrease in fuel delivery to maintain constant τ.

Once the engine is performing well under steady state conditions, it is time to look at transients. During tip in, there is a temporary rush of air to the manifold and cylinders. This spike in delivered airflow must be accompanied by a matching increase in fuel delivery to compensate for the loss of

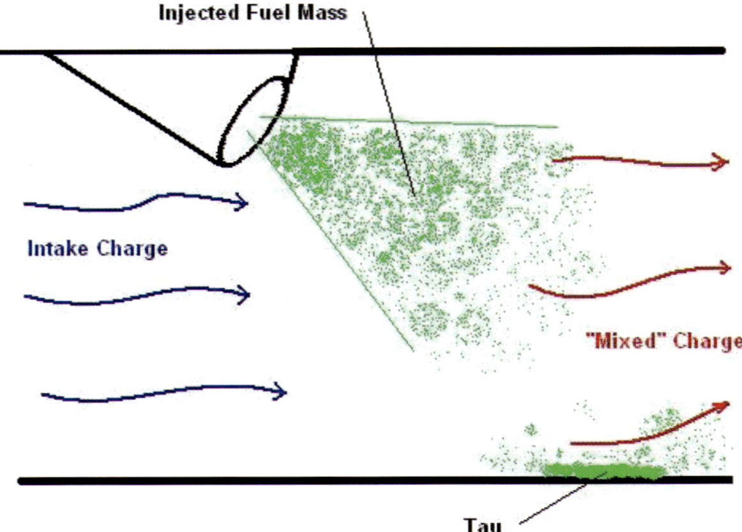

Not all fuel injected into the engine goes directly into the cylinder. A small amount collects on the port walls only to evaporate later. Keeping this amount of evaporation from the walls consistent is the function of the acceleration enrichment logic. (Nate Tovey)

wall film. Without this additional fuel increase, the engine experiences a temporary lean condition that is felt by the driver as a loss of torque or "lag" of power onset followed by a surge to what is normal power for that throttle position. The larger the camshaft or higher the supercharger boost, the more pronounced this effect can be. On a carburetor, the accelerator pump is mechanically actuated by primary blade movement. In EFI systems, a separate "acceleration enrichment" (AE) function is often used to model the necessary temporary fuel increase. The net result of this function is either an adder to the delivered fuel pulse or an additional asynchronous pulse during the following cycle. EFI systems have the benefit of being able to make the AE function variable with respect to throttle position and temperature. Initial throttle opening from closed makes a bigger difference in airflow and requires a larger AE compensation than going from 70 to 100% throttle.

Cooler engine temperatures also require slightly larger amounts of acceleration enrichment.

When the driver lifts completely off the accelerator pedal at higher engine speeds, it is obvious that no engine torque is required. To help slow the engine down, many EFI systems employ "deceleration enleanment (DE). DE is a condition where some or all of the fuel is temporarily shut off to the engine under closed throttle. Without any fuel to burn, the engine is left with pumping losses that exceed the (reduced) power production, causing much quicker deceleration of the crankshaft. The engine speed drop can be harnessed through the transmission to slow the vehicle as well. A side benefit that many OEMs enjoy is that DE reduces fuel consumption. The only trick to properly using DE is the transition back into normal operation. Care must be taken to ensure that normal operation is resumed at a high enough speed to prevent stalling. Additionally, some

hysteresis must be built into the system to prevent rapid changes between normal operation and DE at low throttle cruise conditions. Reducing fuel delivery during light deceleration also prevents excessive buildup of wall film that would otherwise lead to an overly rich condition when returning to part throttle operation.

Pumping pure air through a catalyst that has recently been deposited with unburned fuel during a WOT run can lead to large exothermic reactions taking place, increasing mid-bed temperatures and shortening catalyst life. The closer the catalyst is to the cylinder head, the greater this effect is. Some high output supercharged OEM applications actually employ enrichment during deceleration to improve close coupled catalyst cooling. The richer-than-stoichiometric mixture burns at a lower temperature and excess fuel can continue to cool downstream exhaust components.

Other more complex systems employ an intake manifold model to predict actual air delivery to the cylinder as throttle position changes. Depending on manifold volume, port design, and sensor placement, it can be two to three cycles before instantaneous MAF sensor changes can be processed into effective injector output changes. Although this is practically transparent to drivers, increasingly stringent emissions standards are forcing OEMs to split hairs to find incremental improvements in lambda control. The intake manifold model is constantly anticipating airflow to each individual cylinder on every cycle. This anticipated airflow value can be verified and updated during relatively steady state conditions and is often used as a rationality check for MAF and MAP sensor performance.

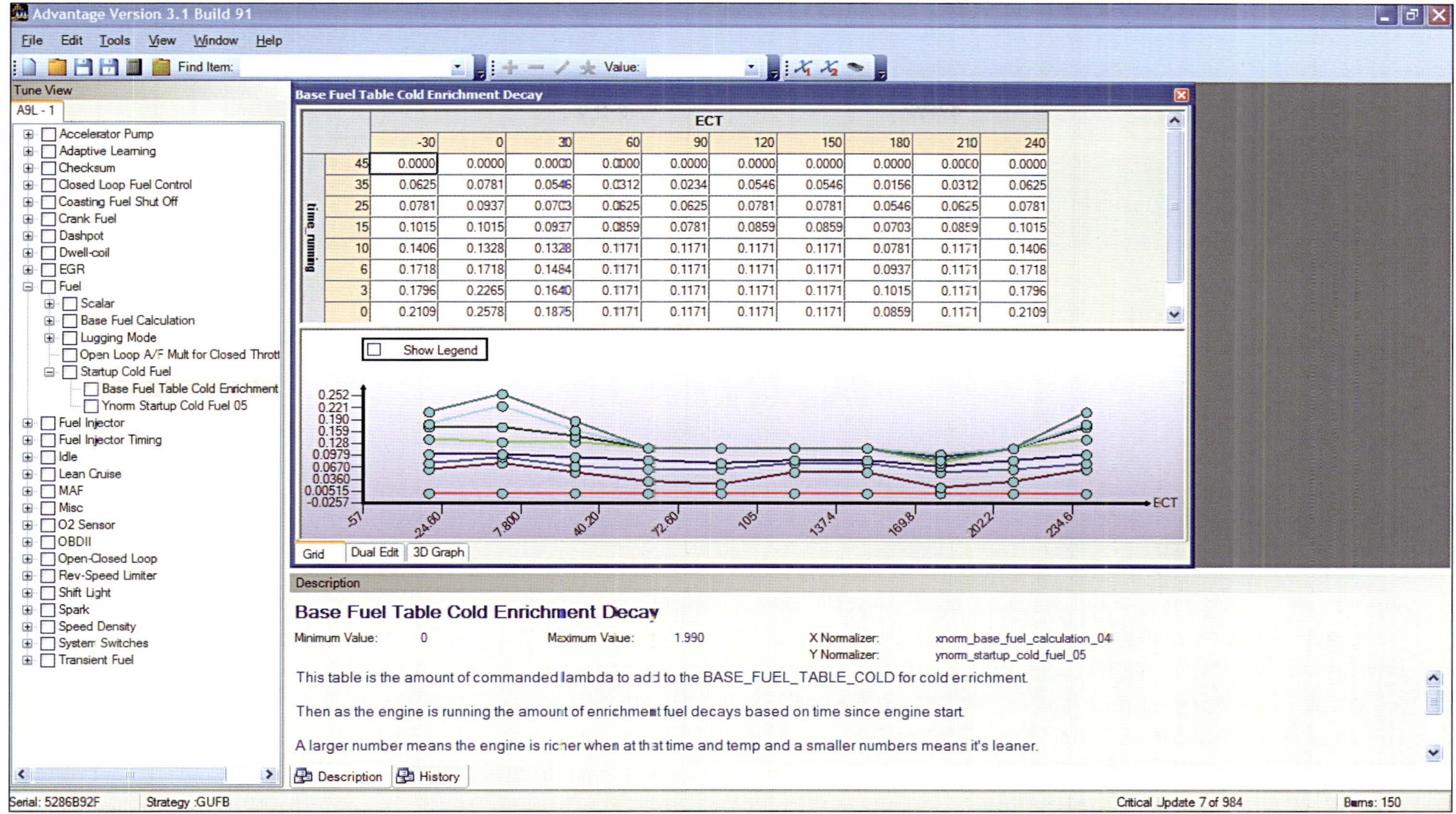

Base Fuel Table Cold Enrichment Decay

time_running \ ECT	-30	0	30	60	90	120	150	180	210	240
45	0.0000	0.0000	0.0000	0.0000	0.0000	0.0000	0.0000	0.0000	0.0000	0.0000
35	0.0625	0.0781	0.0546	0.0312	0.0234	0.0546	0.0546	0.0156	0.0312	0.0625
25	0.0781	0.0937	0.0703	0.0625	0.0625	0.0781	0.0781	0.0546	0.0625	0.0781
15	0.1015	0.1015	0.0937	0.0859	0.0781	0.0859	0.0859	0.0703	0.0859	0.1015
10	0.1406	0.1328	0.1328	0.1171	0.1171	0.1171	0.1171	0.0781	0.1171	0.1406
6	0.1718	0.1718	0.1484	0.1171	0.1171	0.1171	0.1171	0.0937	0.1171	0.1718
3	0.1796	0.2265	0.1640	0.1171	0.1171	0.1171	0.1171	0.1015	0.1171	0.1796
0	0.2109	0.2578	0.1875	0.1171	0.1171	0.1171	0.1171	0.0859	0.1171	0.2109

Extra fuel is added to the engine at cold temperatures to offset the reduced evaporation rate.

Correction Factors

During extreme temperature conditions, it becomes necessary to adjust lambda targets to improve drivability or durability. Extremely cold conditions make fuel evaporation slow and difficult. To make sure that a consistent amount of evaporated fuel is available for combustion, some excess total fuel mass must be added until temperatures normalize. This often means that a significant amount of unburned fuel and HC emissions are passed through the exhaust. Although this is in direct opposition to good emissions performance, the vehicle can sometimes be literally undrivable without some degree of cold enrichment.

Hot ambient conditions reduce the engine's cooling capability and the hotter associated intake charge temperatures are more prone to detonation. Increasing fuel delivery reduces the total amount of heat absorbed by the engine during combustion and can be used in a limited capacity to provide for a "hot limp mode." Since hotter intake charges lead to faster flame speeds, a little extra fuel cooling of the charge reduces chances of detonation at high temperatures. Because of their susceptibility to detonation, forced induction engines often have aggressive fuel compensation at high temperatures.

During startup, the intake manifold and combustion chamber are relatively cold. Much like during cold ambient conditions, fuel evaporation is relatively slow. In order for the engine to run, fuel must be evaporated prior to ignition. If only a portion of the fuel injected into the engine has evaporated, the result will be a lean condition at the time of ignition. EFI systems offset this problem by adding a larger total quantity of fuel to the engine, knowing that only a portion of this fuel delivered actually burns. This extra fueling results in the correct air mass to fuel vapor mass ratio inside the combustion chamber when the engine is cold.

As running time passes, the engine warms up and a larger percentage of the fuel being injected evaporates prior to ignition. The amount of excess startup fuel being added can be decreased to zero. This amount of cranking fuel and startup enrichment should decrease with respect to engine temperature to avoid flooding on hot restarts.

IGNITION

Knowing that engines produce power by harnessing the dramatic pressure increases resulting from the ignition of the air/fuel mix, let's take a look at how that pressure plays into the calibration requirements. Cylinder pressure is directly related to engine efficiency and output. Engineers often refer to BMEP, or brake mean effective pressure, to describe engine operation. Throttling airflow and changing ignition advance have the biggest impact on BMEP, but cam timing, manifold pressure, and lambda also play a large part in determining cylinder pressure. The underlying concept is that higher cylinder pressures generally indicate a more efficiently running engine under normal operation. More efficient combustion requires less ignition lead. If the calibrator can recognize where the engine is most efficient, it becomes easier to determine ignition needs.

If one were to look at a map of actual engine efficiency and compare this to the ignition map showing MBT timing, the relationship between cylinder pressure and required advance becomes evident. Conversely, we can employ this knowledge to use ignition advance to increase cylinder pressure (and torque) in areas where the engine is less efficient. Advancing ignition timing at low RPM under load can help improve throttle response before the camshaft and intake runner efficiency come up in the midrange. Likewise, at higher RPM after the engine passes the peak efficiency range of the cam and intake runners, cylinder pressure begins to fall off. More power can be made by increasing ignition advance to make up for the decreasing engine efficiency as the engine approaches redline.

Calibrators must be keenly aware of two unique conditions of increased

Ready for another routine 9-second run, this Mustang warms the slicks in an attempt to plant all of its power. (Nate Tovey)

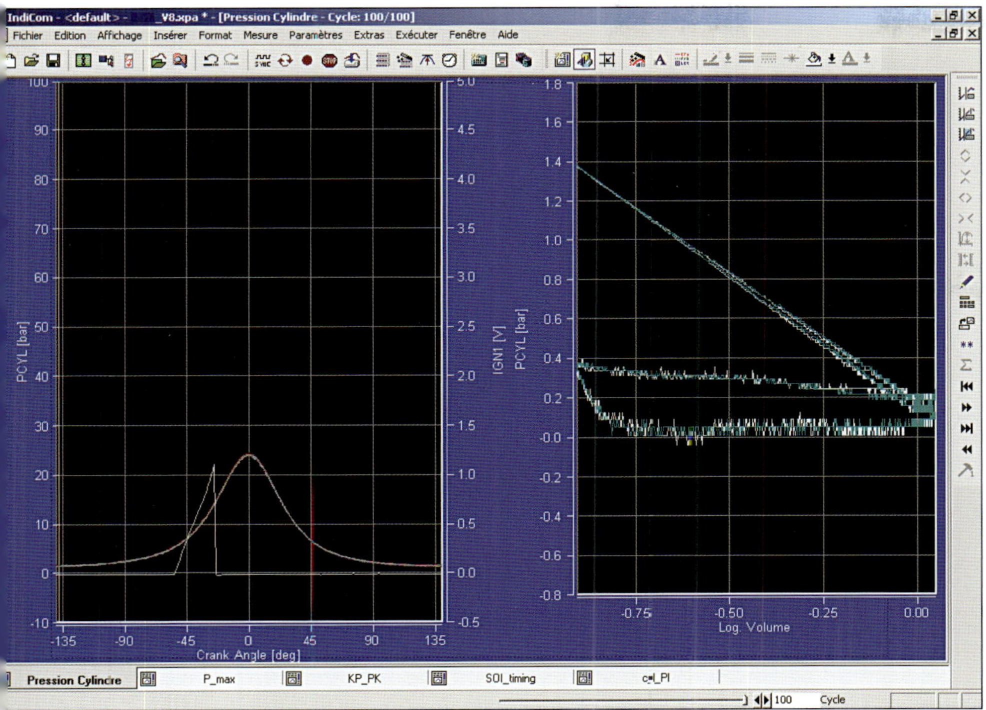

Using the AVL Indicom software and pressure transducers in the cylinders, we can see both the P-V diagram (right) and pressure versus time (left) for motoring the engine. The triangular spike on the left is the spark event.

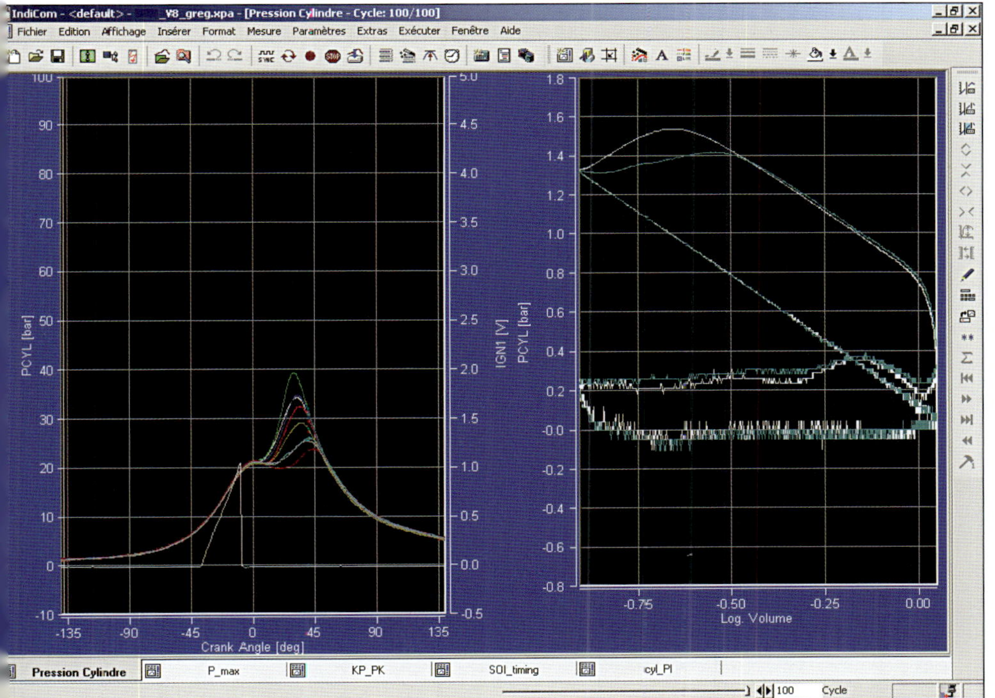

Here we see the pressure traces for a running engine with less than ideal spark timing. Notice how the pressure rise happens very late in the cycle and each cylinder has its own unique trace line.

cylinder pressure. Knock and preignition have very similar symptoms and results with slightly different causes. Knock is caused by an uncontrolled fast burn in the chamber after the initial spark event. This is often the result of excess ignition advance, lean mixtures, or elevated temperatures. Preignition is the premature start of combustion before the normal spark event. This is often caused by hot spots in the combustion chamber or residual glowing hydrocarbon deposits that ignite the mixture before the spark plug has a chance to fire. Either way, both conditions mean that peak combustion pressure has been reached before the piston arrives at TDC. This means that not only is normal combustion pressing on the cylinder head, gasket, and piston, but dynamic compression happens as well. The symptom for this condition is usually a sound similar to marbles in a coffee can. While not always audible, the intensity of this noise grows with the intensity of engine stress resulting from the condition. The forces associated with knock and preignition can be as much as two or three times the normal combustion forces. This can be incredibly destructive to engine components, often resulting in a blown head gasket, broken piston, bent connecting rod, or failed bearings. Knock should be avoided, period.

Burn Rate

Burn rate of the air/fuel mixture has a dominant impact upon just how much ignition lead an engine wants. The speed at which the combustion process occurs directly influences the rate of cylinder pressure rise.

Most spark ignition engines have an effective flame propagation speed of 706 to 985 in/s (18 to 25m/s). This means that a flame starting from a central spark plug in a 4.00-inch bore engine can reach the cylinder wall in about two and a half milliseconds. At an idle speed of 800 rpm, this translates into roughly 11.5 degrees of crankshaft revolution. That same flame speed and bore size at 3,000 rpm yields more than 43 degrees of crankshaft rotation during our theoretical burn time to reach the chamber walls. This is why ignition timing must advance at higher speeds to ensure combustion before the piston has a chance to run away from the reaction.

Actual flame propagation speed changes with conditions. Faster flame speeds and better charge mixing during combustion speeds up the reaction that is producing the expanding gases and rising pressure. Burn rate is influenced by a number of engine operating conditions beyond just chamber design.

Fuel octane is a number representing the ratio of octane to heptane in gasoline. Higher octane numbers indicate a stronger concentration of the slower, more stable- burning octane molecules than the more volatile heptane. Since high-octane fuels have a slower burn rate, larger amounts of ignition lead can and should be used. Increasing the ignition lead results in an increase in torque, so higher fuel octane levels allow for more power resulting from the necessary ignition advance increase. Keep in mind that fuel octane is a global change to the engine's ignition advance needs. Running higher octane fuel means more spark advance at idle and part load as well as WOT. Changing from 93 octane pump fuel to 110 octane racing fuel may allow 4

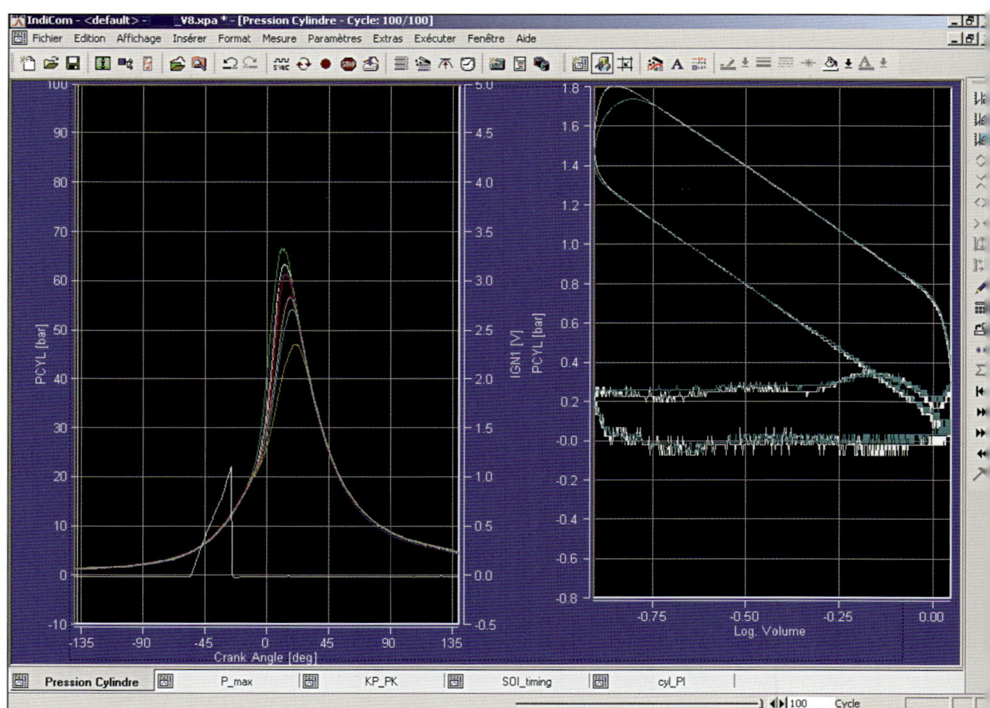

Figure 7-1 *Pressure trace at part load with ignition adjusted to MBT. Cylinder pressure is maximized just after TDC and area under the curve in the top portion of the P-V diagram is also greatest.*

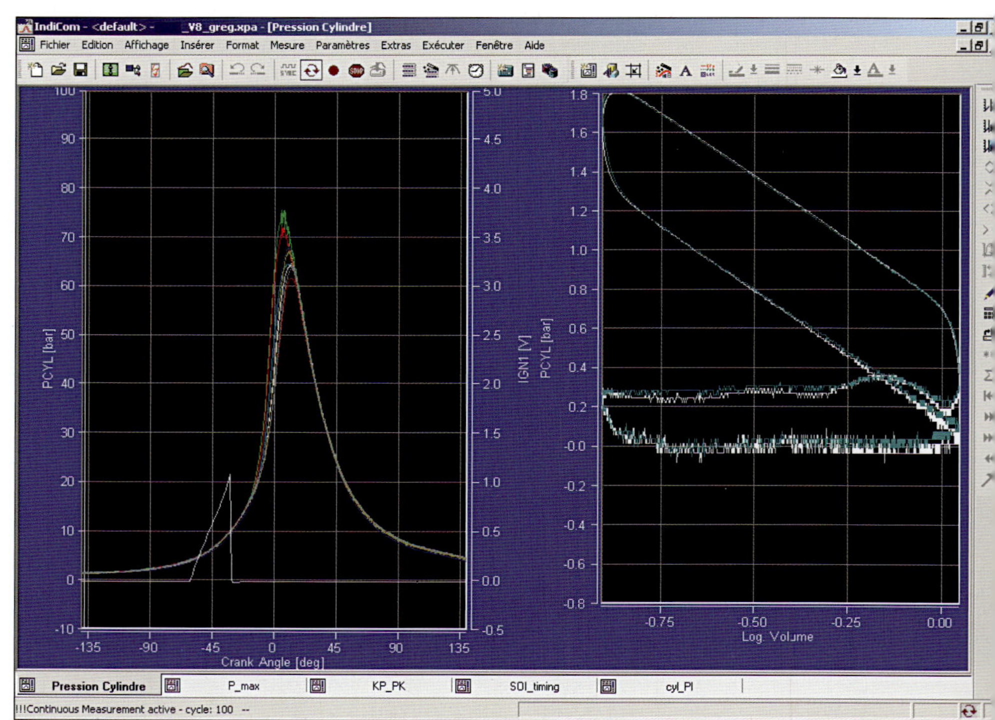

Figure 7-2 *Pressure trace showing knock resulting from a spark event triggered about 6 degrees before MBT at part load. Notice the ragged tips of the pressure traces and increased cylinder-to-cylinder variation.*

degrees of additional spark advance at WOT (with an accompanying increase in power), but may idle poorly and exhibit reduced fuel economy and throttle response unless all other spark tables are increased by a like amount.

Lambda changes burn rate, often requiring an adjustment to commanded ignition advance. Slightly rich ($\lambda \approx 0.9$) mixes tend to burn quickest, requiring less ignition advance to reach peak cylinder pressure at the right time. Base and MBT ignition tables are usually calculated for $\lambda = 1$ for OEM applications with a second table to adjust for changes in lambda. These tables are typically a global ignition adder to increase ignition advance whenever an excessively rich condition is entered. This helps compensate for the changes in burn rate as well as keep exhaust gas temperatures within an acceptable range.

Charge temperature has a similar effect on burn rate. Hotter intake charges more readily evaporate fuel and increase the burn rate. Much like lambda, the OEM solution usually involves an adder function to reduce global timing with higher inlet temperatures. For this to work properly, the IAT sensor must be installed

Cylinder Pressure Measurement

So you're ready to calibrate like an OEM engineer? We've spent quite a bit of time in this book discussing cylinder pressure and the factors that influence it. We've even covered BMEP and how to use it for a valid comparison. So how much pressure is really in the cylinder and how does one measure it?

The most common answer is a pressure transducer located somewhere in the cylinder itself. As one might suspect, fitting a pressure gauge inside of a running engine is a bit of a tight squeeze. Precious few openings are available, and those that are available are all occupied. The solution becomes a combination of pressure transducer and a normal engine component. Spark plugs are the easiest targets for such combinations due to their ease of access from outside of the engine. Specially modified spark plugs are made to include a piezoelectric pressure-sending unit next to the electrode. There are also some variants that have a pressure transducer integrated in a head gasket. The signal from this sending unit can be fed to an oscilloscope or high-speed data acquisition unit for processing.

So what does this data really tell the calibrator? First and foremost, it shows not only whether combustion has occurred but at what quality. This translates into misfire measurement when pressure rises fail to exceed the normal pumping function of the engine. Additionally, the size and shape of the curve is an indicator of power being generated by the cylinder. Locating this curve with respect to TDC helps identify the effectiveness of spark advance.

Knock will also show up on cylinder pressure traces. The normal trace is a relatively smooth bell curve (**Figure 7-1**). When knock is present, the top of this curve (where pressure is highest) exhibits increasing roughness. The abnormal combustion associated with knock shows up as the inconsistent pressure trace (**Figure 7-2**). A significant advantage to using cylinder pressure measurements during the calibration of an engine is that the pressure variations associated with knock will show up on such a reading long before the condition can be detected with a conventional knock sensor. Many OEM engines are now calibrated to operate with a certain tolerance of exactly "how much knock is acceptable."

A common rule of thumb describes the allowable knock limit expressed in terms of maximum cylinder pressure versus average peak cylinder pressure and RPM.

Allowable Over-Pressure (bar) = RPM / 1000

This means that any amount of knock resulting in more than 4 bar additional cylinder pressure at 4,000 rpm would be unacceptable, and possibly engine damaging.

Time and cost restraints often mean that cylinder pressure monitoring is not a viable option for the casual tuner. However, if you plan to work on many examples of the same engine and would like to find out exactly where the knock limit lies without overstepping into the danger zone, such a setup may be well worth the money and complexity.

such that its output is directly proportional to actual inlet temperature to the cylinder. Installation of a supercharger or large amount of plumbing between the IAT sensor and the intake port skews the reading and reduces accuracy of this compensation curve. If the IAT is installed after a supercharger, it has the benefit of allowing the PCM to adjust timing to compensate for heat soak in the compressor, improving safety margin against detonation at high temperatures.

Engine load is an indicator of final charge density in the combustion chamber and ultimately reflects engine efficiency. This charge density increase can come from an ambient barometric condition, throttle position, artificial increase in manifold pressure (supercharging), or higher compression ratio. The result is a mixture of more closely packed molecules that allow the flame front to easily hop from one molecule to the next in the chamber. These more efficiently packed molecules burning at a faster rate again require less ignition timing advance to reach peak pressure at the proper moment.

Mechanical mixing of the air/fuel charge inside the cylinder helps to increase bulk burn rate. Although actual flame speed remains relatively constant, the tumble and swirl motions inside the chamber move the mass of burning charge around, exposing it to unburned mass even more quickly. The result is a faster, more complete combustion event.

"Balancing the Players"

When it comes to increasing or maintaining cylinder pressure, we now see multiple ways to get to the desired output. If an engine can only tolerate a given amount of cylinder pressure, the calibrator often finds himself with several options to achieve that pressure. There is a balance that needs to be preserved between lambda, spark advance, load, and temperatures. This is especially true at WOT where the same power may be developed using a lean mixture with less timing or a rich mixture with increased advance. It is

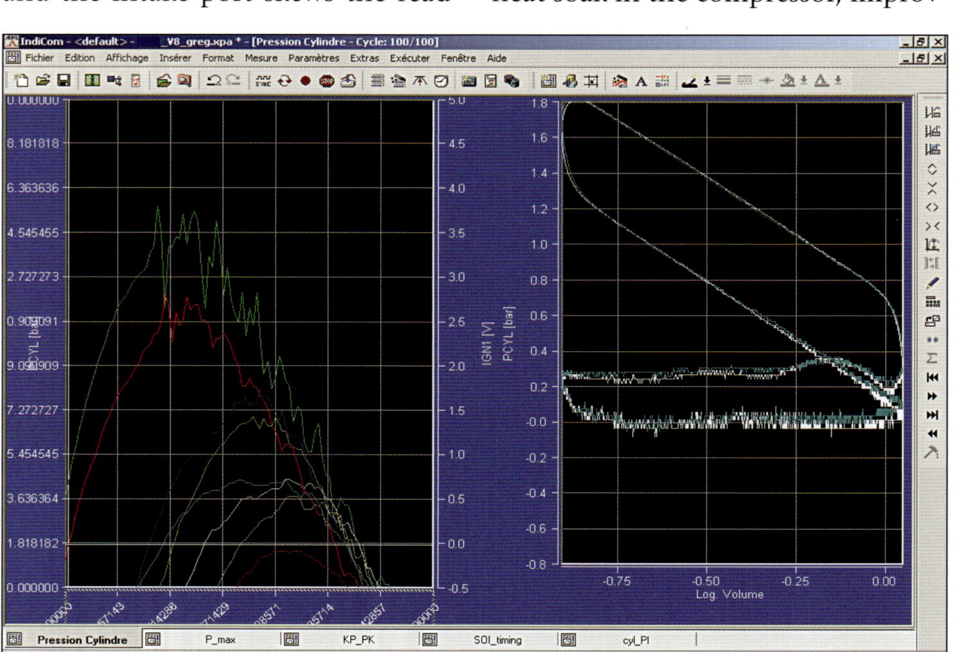

A closer look at the pressure trace shows the erratic nature of the cylinder pressure during knock.

The Subaru STI features one of the most advanced engines today. Electronic throttle control, active knock control, and an intercooled turbocharger help contribute to a specific output of over 120 hp per liter, with plenty of room for aftermarket improvement.

up to the calibrator to decide which balance to strike. There is no free lunch at the limit. If an engine has reached its knock limit at $\lambda = 0.82$ with 26 degrees of advance, we must either increase fueling to add ignition lead or reduce timing to find power with a leaner mix.

At cruise, the balance is simpler because we are usually targeting $\lambda = 1$. In this case we simply find the ignition lead that develops MBT (highest thermal efficiency) and allow for adjustment for environmental changes. Since the engine is operating in a range where flame speed is slightly slower than under full power, some additional timing must be added to optimize combustion. Many new calibrators are surprised by just how much ignition advance an engine requires for best operation at light load and cruise conditions compared to full power operation.

MBT and Mass Fraction Burned

It is important to understand that the fuel and air inside of the cylinder do not instantaneously react. It is a process that happens over time, however short this time really is. The rise of pressure inside of the cylinder is proportional to the amount of combustion that has taken place as well as the piston's position. If we take a look at how much of the charge has been burned relative to piston movement, we begin to understand why the pressure peaks when it does. (**Figure 7-3**). Since the start of combustion is controlled by ignition timing advance, the relationship between ignition timing and cylinder pressure shows its connection to what engineers call X_b, or "mass fraction burned." The object is to first attempt to achieve close to 100% X_b prior to the opening of the exhaust valve. This helps to harness much of the expansion energy during the power stroke. The second and more important point is to control the pressure rise relative to normal compression.

If combustion is timed to release peak pressure at a point where the engine can mechanically take best advantage, the largest possible torque is generated at the crankshaft. This peak mechanical efficiency is usually between 7 and 15 degrees ATDC on most piston engines.

By aligning the point at which 50% of the mass fraction has burned with the point at which the piston and connecting rod have a maximum mechanical acvantage on the crankshaft, one can work backward to find an ideal ignition point for those specific conditions. Not surprisingly, this ignition lead point is referred to as MBT (maximum brake torque), as it delivers the highest possible engine torque for the conditions.

OEM engineers spend large amounts of time doing computer modeling of the combustion process to identify the mass fraction burned curve of their engines. This information is then used to back calculate to a theoretical MBT timing map before the engine is ever started. In testing, engineers measure cylinder pressure versus crank angle to determine the rate of heat release. The point at which 50% of this heat release has occurred is known as "CA50," or "crank angle for 50% burn." A slight delay occurs between heat release and pressure build, so engineers aim to achieve CA50 a few degrees ahead of peak pressure. The result is that a CA50 of 5 to 12 degrees ATDC correlates with ignition timing equal to actual MBT.

For the aftermarket calibrator this number is less important since the MBT table is usually found by empirical dynamometer testing in the absence of expensive and time consuming engine modeling software or in-cylinder pressure measurements. Knowing where it comes from gives an insight to exactly what happens inside the cylinder when spark tables are adjusted.

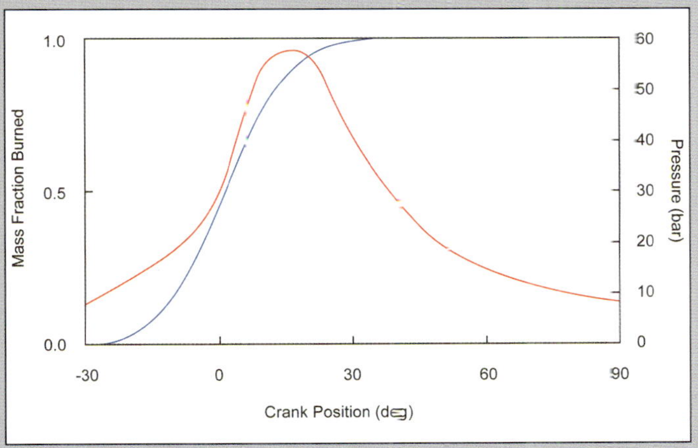

Figure 7-3 *Mass fraction burned (blue) and cylinder pressure (red) versus crank angle. Notice the slight delay between 50% burn and peak pressure. (Nate Tovey)*

DATA LOGGING

Before we begin changing PCM parameters, let's take a look at the key elements to monitor. Most tables and functions in the PCM are constructed with axes that represent a range of engine operating conditions. In order to know where to make changes to the PCM tables, it is necessary to know exact engine conditions at any time. This data may be available in real time from instruments or a computer link to the PCM, or recorded in a datalogger for later review.

Many PCM programming tools (including almost all aftermarket PCMs) allow for real-time changes while the engine is running. This is the ideal method of calibration for most parameters since it allows the calibrator to see exactly where in the tables the engine is operating. If the PCM does not support real-time adjustment, the next best thing is real-time monitoring by way of some scan tool. Multiple OBD-II capable data readers are available, such as EFI-Live, AutoTap, SCT Raptor, Diablo Predator, and OEM tools such as the GM Tech2, Ford NGS, and WDS. These tools also allow the user to record brief periods of time, which is

The GM Tech2 is a dealership-level diagnostic tool that can record additional data parameters not available to basic scan tools. It can also serve to perform vehicle specific diagnostics and limited programming duties. Ford and Chrysler have similar tools called the NGS and StarScan, respectively.

Aftermarket tools like the Diablosport Predator perform double duty as programming and datalogging tools. (Nate Tovey)

extremely useful when measuring transients and WOT behavior where it is impossible to plot one cell at a time on the tables due to fast engine sweep rates.

At a minimum, the calibrator needs to see real-time displays of RPM, load, ECT, IAT, and lambda to calibrate most base tables. Load may be shown as MAF or MAP output or calculated load. It is easiest to view these in real time while working on a dynamometer to facilitate holding at a fixed speed and load point for calibration.

Engine coolant and air temperatures should always be monitored to ensure that we are working in the normal operating range. We can always go back and adjust for cold start or heat soak later. Not confirming where the engine is operating on a correction table begs for trouble.

Actual fuel delivery should be monitored at low load to check for proper fueling. Excessively large pulsewidths at idle may indicate faults somewhere else. This may point you to a problem with a sensor or correction curve that is nowhere near correct. Additionally, WOT duty cycle should be verified to ensure that the fuel injectors are not going static, forcing a lean condition. If this is the case, it becomes impossible to correct WOT mixture by anything other than a pressure or injector size increase.

Ignition timing should be monitored closely at idle and WOT. An unstable amount of ignition lead at idle often leads to unstable fuel mixes. Many hours have been wasted by tuners attempting to fix a timing problem with the fuel trims. It is not uncommon to see 10 degrees of swing in idle timing to correct for fluctuations in speed, but 20 to 30 degree changes usually indicate a load measurement or speed pickup accuracy issue. Actual ignition timing at WOT should be recorded for review after a dynamometer pull or drive. Since the engine sweeps rapidly through the RPM range at WOT, it is best to review this data after coming to a stop. This also gives the calibrator time to review any indicated knock sensor input or temperature related spark retard at the same time.

When tuning idle speed and mixture, it is important to monitor

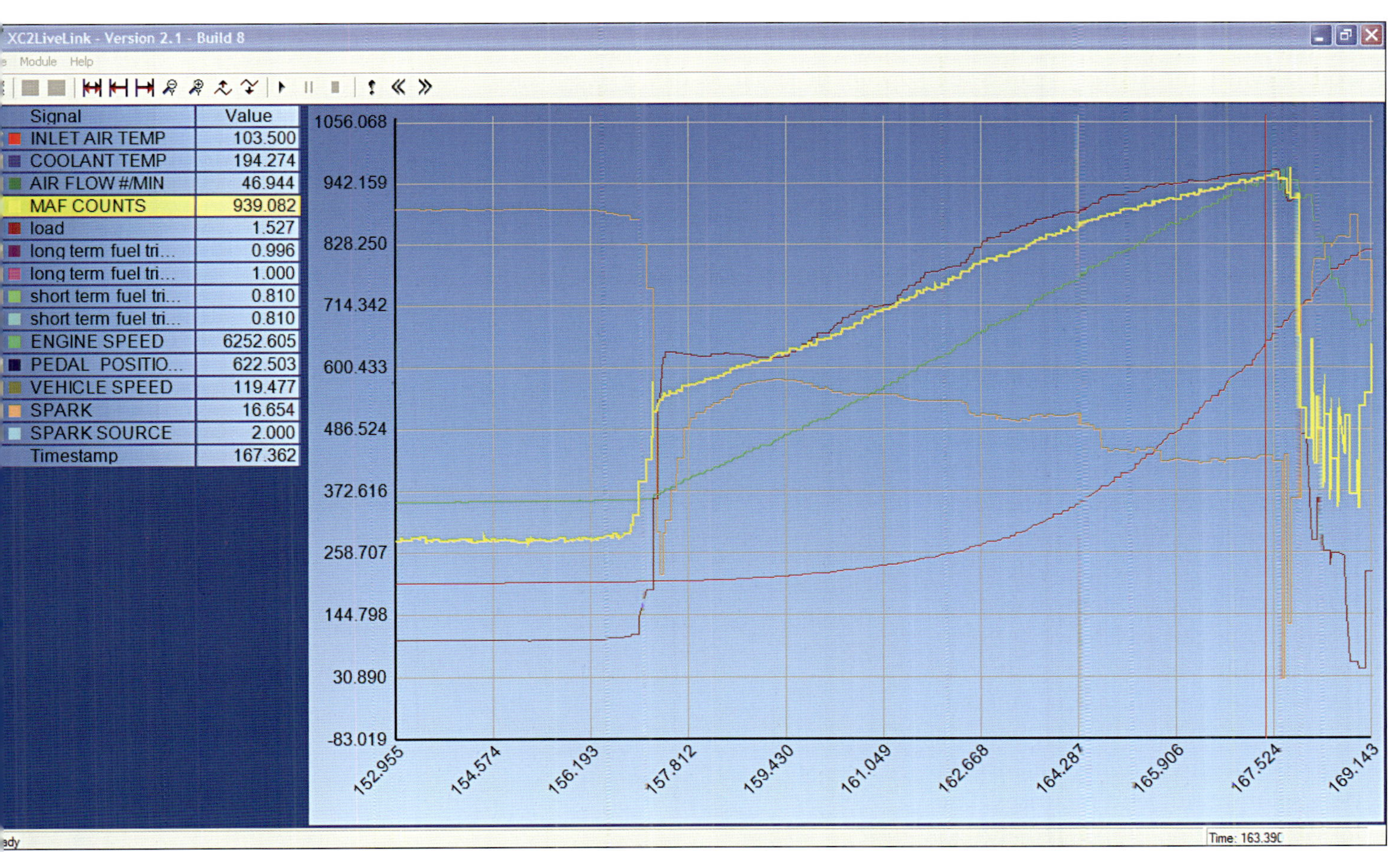

During the tuning process, it is important to record many parameters at once. The Livelink software from SCT shown here includes a simple interface that can be configured to the user's preference.

IAC motor position or duty cycle. This is the fine adjustment to actual idle speed, and the calibrator should ensure that it is actually being used. If the IAC displays as either all the way open or closed at all times, it's time for a throttle stop adjustment. The more range the IAC motor has in both directions, the better chances it has of controlling a stable idle speed.

PCM fueling adjustments should also be closely checked at idle and part load. If the engine has entered closed loop operation, it uses the signal from the HEGO to trim fuel delivery back to the target ($\lambda = 1$). If the engine has not entered closed loop, the calibrator can safely assume that the base fuel tables are determining actual fuel delivery. In open loop, the base fuel calculations in the PCM can be adjusted by referencing actual λ from a wideband sensor against the desired air/fuel ratio from the PCM tables. If a closed loop operation is functional, the PCM's fuel trims can be used to determine the error in the base fuel calculations.

Know Your Load

Engine loading is not the same for all vehicles. The intended use of the vehicle dictates the required calibration. Vehicle loads change the actual engine load greatly depending upon how the vehicle is used. A drag racer taxes the engine differently from a top speed racer, boat, or road racer. The largest difference between the definitions of WOT usage is temperatures and stabilization. Durability and temperature control become serious concerns as vehicle loads and time spent at full load increases. The longer the engine is operated at high loads, the more heat is transferred from the combustion chamber to the

Drag racing loads the vehicle very heavily for a relatively short amount of time. (Nate Tovey)

valves, cylinder head, coolant, oil, under hood area, exhaust system, and catalyst.

Due to its short duration, drag racing is a remarkably forgiving environment as far as racing goes. Races are completed in a matter of seconds. The vehicle load at the start is only its own weight. This load increases steadily toward the end of the track as aerodynamic forces increase resistance on the car. Because heat transfer takes time and the races are short, a lower amount of total heat is imparted to engine components in a drag race compared to other forms of motorsports. Extended cool-down periods between races further reduce temperature issues for serious drag racers. This means that engines can be calibrated to operate at slightly more aggressive power levels, generating more heat during these brief tests without overheating.

Road racing is a different environment for an engine. Road racing

Road racing gives the engine plenty of time to heat soak while still being asked to repeatedly deliver full power in bursts.

requires the car to operate at changing loads, often very high, for extended time. Much of this racing happens at high speeds, so aerodynamic loads are significant for most of the operation time. During the course of a race, it can be expected that oil, coolant, and under-hood components all eventually come to a stabilized, elevated temperature. To aid durability, exhaust gas temperatures should be considered strongly when performing WOT calibration. It may be necessary to sacrifice some power to reduce temperatures by running a slightly rich air/fuel ratio. Spark advance should generally be kept close to optimal since retarded timing increases exhaust temperatures. This becomes even more critical on street-legal vehicles that use a catalytic converter. Excessive exhaust temperatures can quickly destroy the monolith of the catalyst, destroying emissions and reducing power as the exhaust gets plugged by the melted substrate. Radiators and cooling systems for these cars often need to be upgraded to prevent the ECU from prematurely pulling power to protect against excessive temperatures.

Marine and top speed racers share the most arduous engine conditions. These vehicles are rarely equipped with catalytic converters or restrictive exhaust systems. However, vehicle loads from aerodynamic drag at high speed or fluid drag on the hull force the engine to work extremely hard for extended periods of time without relent. Without proper cooling, these engines would self-destruct from constant heating of the combustion chambers. Primary cooling comes from the radiator or lake water feed to the cooling system, but piston and chamber temperatures must also be controlled by adjusting burn temperature. Running a rich mixture with a maximum of tolerable spark advance is usually required to keep temperatures safe.

Calibrators use dynamometers to recreate engine-loading conditions in a controlled environment. Not all dynamometers are created equal, and some have significant advantages to the calibrator.

Engine dynamometers have been a long-time standard for both tuners and OEMs. Engine dynamometers attach the engine to a variable

The chassis dyno allows the car to be driven in place while measuring power and torque. This controlled environment gives more repeatable results and an immediate indication if tuning changes were effective. (Nate Tovey)

electric or hydraulic load source. Speed and load can be independently adjusted, and a controlled sweep can also be done. They allow for terrific control over coolant, oil, and inlet air temperatures. This takes a lot of the guesswork out of calibrating the base maps and makes it easier to build temperature compensation tables later. The disadvantage is that the engine dynamometer rarely employs the same exact exhaust system as the vehicle. There can be as much as 20% difference in an engine's volumetric efficiency between operation on an engine dynamometer with open headers and in a vehicle with catalysts, mufflers, and tailpipes. Speed density systems need to be calibrated with the exact hardware combination they use in the vehicle to prevent a mismatch.

Chassis dynamometers allow the entire vehicle to be tested by placing the wheels on rollers and

The engine dynamometer allows the tuner to focus strictly on the engine's performance, but tuning should always be double-checked in the vehicle. (Nate Tovey)

The inertial chassis dyno is perhaps the most common dyno used today. Low cost and easy operation make them an attractive tool to the aftermarket. (Nate Tovey)

measuring output. Two main types of chassis dynamometer exist: inertial and load bearing. Each has its own advantages.

The inertial chassis dynamometer is a simple design that uses a weighted drum with a speed pickup. With a known drum mass, the acceleration rate can be used to determine power input. Vehicle torque output is extrapolated based upon speed. While this method is very consistent from run to run, the loading on the vehicle remains constant regardless of speed. Inertial dynamometers are very useful for "A" to "B" comparisons to see if a change to a vehicle increased engine output or not. The low cost of the inertial chassis dynamometer makes it a popular choice for many performance shops looking to show customers power increases from their purchases.

The load bearing chassis dynamometer is more complex and costly. The same rollers are fitted with a load absorption unit that is controlled by a computer. Actual vehicle output torque can be directly measured rather than extrapolated. In the real world, vehicles experience increasing loads from aerodynamic drag as speed increases. Inertia alone cannot give a true representation of vehicle loading on the street. The load bearing dynamometer has the ability to increase load with speed to accurately simulate actual vehicle loads. The dynamometer can be programmed to give different aerodynamic drag loads to a pickup truck versus a sports car. Additionally, the load absorbers can be used to hold the vehicle at a fixed speed or loading point, easing calibration work. Turbocharged vehicles in particular benefit from calibration on a load bearing dynamometer since engine load dictates how quickly the turbocharger responds.

Both the inertial and load bearing chassis dynamometers share a common drawback. Since the vehicle is actually stationary while the wheels

Load bearing chassis dynamometers like this MD1100 allow the calibrator to test at steady state as well as realistic dynamic sweeping conditions. Cooling the vehicle during these tests often requires several external fans. (Nate Tovey)

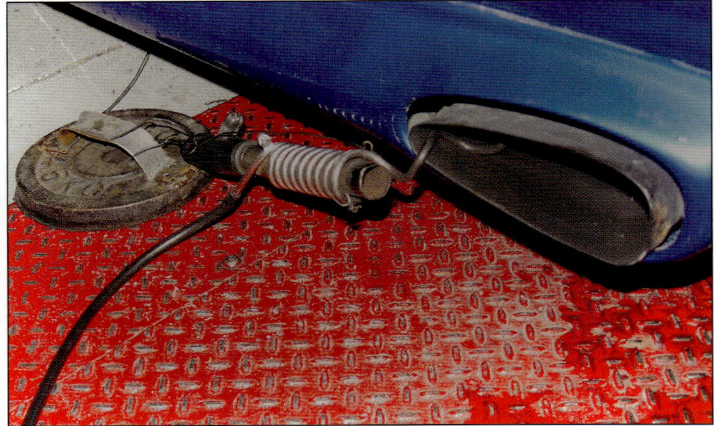

Many chassis dynos use a "tailpipe sniffer" to monitor air/fuel ratio. This is at best inaccurate and lagging, and in many cases may yield deceptive results. When using a wideband, it's always best to install it properly closer to the engine. (Nate Tovey)

move, airflow across the vehicle (particularly the cooling components) is not the same as seen on the street or track. While most dynamometers are fitted with cooling fans, few can replicate the sheer volume of airflow encountered in the real world. Engine temperatures should always be closely monitored to allow for proper cooling. If the vehicle is equipped with any other heat exchanger (such as an intercooler) charge temperatures on the dynamometer will almost certainly vary compared to real world operation. The dynamometer should be used to develop the majority of the calibration, but there is no substitute for real world testing at the end of the project.

NTK AFX Wideband

As a calibrator used to working with lab-grade equipment, it's easy to become spoiled with the quality of data one can collect. When it comes to aftermarket data collection there are almost always significant compromises in this arena. Nowhere is this more prevalent than the choices for wideband oxygen sensors. For years, the only ones available were the lab grade instruments used by OEMs in their development. The cost of such units made them all but unavailable to anyone except the most dedicated of aftermarket calibrators. Until recently, somewhere between $3,000 and $10,000 was needed to get a wideband.

There are now many choices on the market for the casual tuner or enthusiast looking to measure air/fuel ratio. All of these units use a sending unit from either Bosch or NTK along with a controller and signal conditioner of their own design. Obviously, some of these units are more accurate than others.

ECM (Engine Control and Monitoring) manufactures some of the most accurate and widely used OEM level lab-grade wideband sensors in the world. They have been historically expensive (especially compared to current aftermarket offerings), but worth every penny to the engineer requiring accuracy inside of one percent. ECM has recently teamed up with Bosch and NTK to offer a new, affordable wideband to the aftermarket enthusiast. Branded under the "NTK AFX" name, the unit is very similar to the much more expensive ECM AFM 600 controller, and shares its accuracy where it counts. The AFX wideband can use either sending unit. It can also be calibrated in free air, has a 0 to 5v analog output for easy integration into any other datalogging system, and a convenient LCD display. All of this functionality is packaged in a compact box that can be easily installed in any vehicle. The AFX is available through several major resellers for a price competitive with any other wideband unit but without the regular sensor failures that result from inadequate control strategies. More importantly, the AFX generates an accurate signal when and where it counts for the aftermarket calibrator.

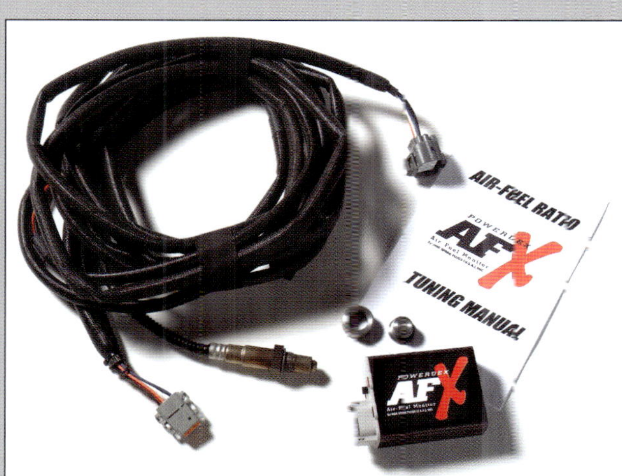

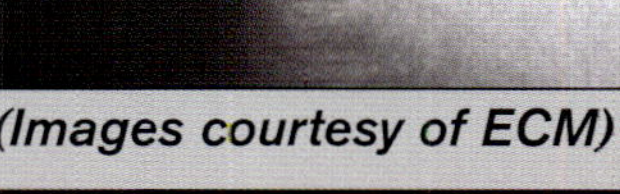
(Images courtesy of ECM)

GETTING INTO THE ZIP CODE

Now that we have a good idea of what an engine needs to operate smoothly, it is time to actually begin to construct the framework of the calibration. Before the engine is even started, the calibrator must provide the PCM with some parameters that are close enough to the optimum setting to begin checking operation. "Getting into the zip code" of the actual ideal settings can be one of the toughest tasks in the whole process. Whenever possible, it is always easiest to start with a known good calibration for the engine combination

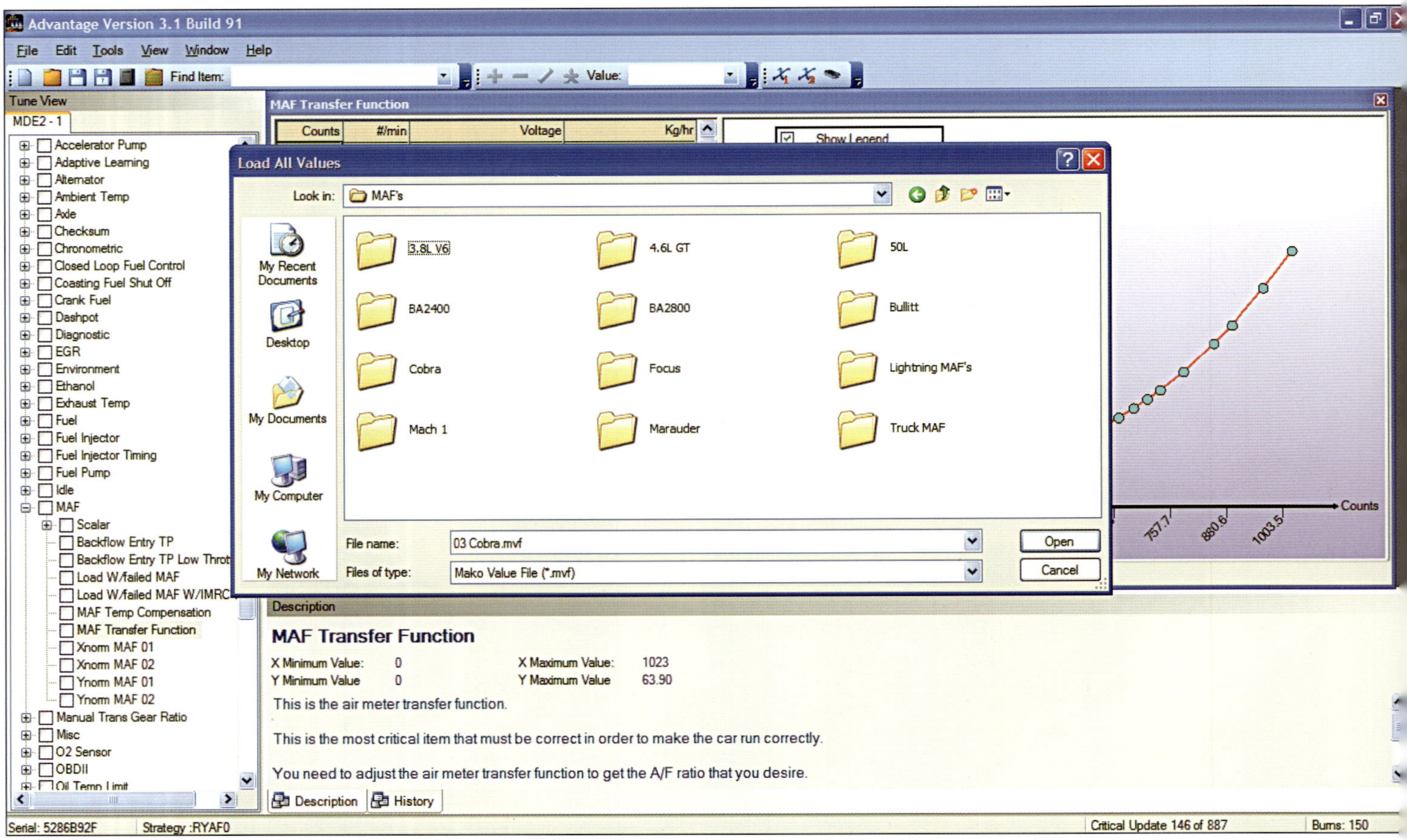

Some software packages allow the calibrator a head start with the necessary changes by loading predefined tables. The SCT Advantage package shown here allows the user to select from a list of MAF sensors as a starting point for the airpath model.

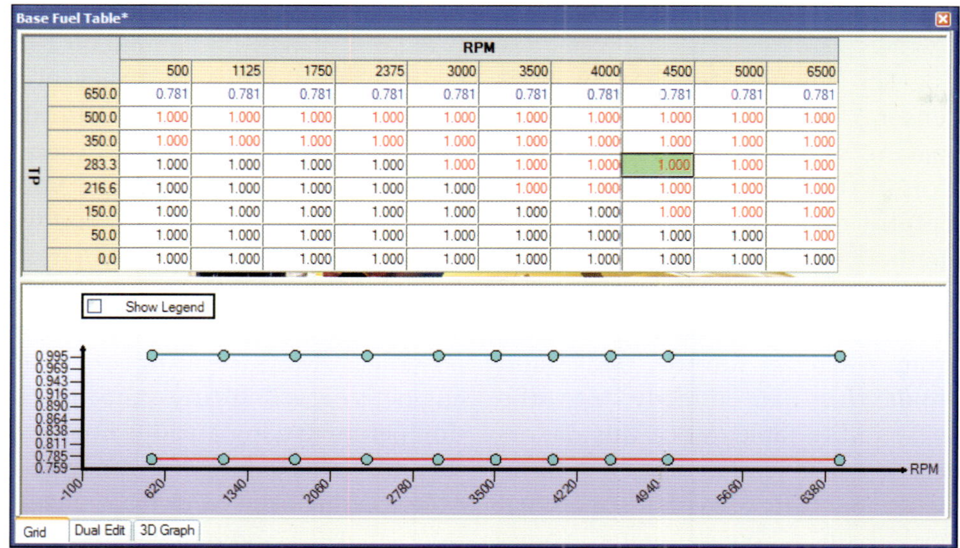

Base Fuel Table*										
	RPM									
TP	500	1125	1750	2375	3000	3500	4000	4500	5000	6500
650.0	0.781	0.781	0.781	0.781	0.781	0.781	0.781	0.781	0.781	0.781
500.0	1.000	1.000	1.000	1.000	1.000	1.000	1.000	1.000	1.000	1.000
350.0	1.000	1.000	1.000	1.000	1.000	1.000	1.000	1.000	1.000	1.000
283.3	1.000	1.000	1.000	1.000	1.000	1.000	1.000	1.000	1.000	1.000
216.6	1.000	1.000	1.000	1.000	1.000	1.000	1.000	1.000	1.000	1.000
150.0	1.000	1.000	1.000	1.000	1.000	1.000	1.000	1.000	1.000	1.000
50.0	1.000	1.000	1.000	1.000	1.000	1.000	1.000	1.000	1.000	1.000
0.0	1.000	1.000	1.000	1.000	1.000	1.000	1.000	1.000	1.000	1.000

Grid | Dual Edit | 3D Graph

The target air/fuel ratio is set to a constant value in all areas where the engine may operate during initial tuning. This helps reduce the confusion associated with trying to hit a moving target when correcting the airflow model.

being tuned. If a similar engine combination has been calibrated before, it saves a lot of time in the following steps. Making small changes to an existing tuning file also reduces the chances for wholesale miscalculations and no-start, no-run conditions.

Some aftermarket tuning packages have built-in tools that can often generate maps based on engine component specs (port size, cam specs, compression ratio, etc.) that are close enough to the optimum setting to get started. Other programs allow for changes to be made to a starting OEM file to accommodate hardware changes (new injectors, different MAF sensor, addition of a supercharger) providing the tuner with a closer estimation of what the engine requires in the calibration. This is also the time to review all sensors to be used on the engine to ensure that their outputs are properly recognized by the PCM. A quick key-on (turning the ignition to power up the PCM) to compare ECT, IAT, and MAP values against ambient conditions prevents confusion after startup.

Fuel injector parameters should also be checked at this point. Make sure that fuel pressure has been set properly and injector size parameters in the PCM reflect the components used and current rail pressure. This is also the time to double-check the voltage compensation curves and dead-time characteristics of the injectors being used. Proper modeling of actual fuel delivery makes later modeling of airflow much easier.

One of the primary keys to properly calibrating any EFI system is the accurate modeling of engine airflow. This is the most important point to remember. The majority of all fuel and spark calculations is based upon airflow or engine load. If the PCM does not know how much air is passing through the engine at any time, it has little hope of accurately metering fuel or controlling ideal spark advance.

To begin calibrating airflow, it is usually easiest to start above idle, often 1,500 rpm or more. Idle is one of the most difficult conditions for

the PCM to control, so it is easier to skip right past this to get things "in the zip code." Running the engine at slightly elevated speed decreases the chances of stalling as a result of less than optimal air/fuel ratio or spark advance. When first starting the airflow modeling process, it may also be helpful to lock the timing to aid stability. The goal here is to build maps representing what the engine does under stable operation. We will return later to smooth out transitions. To accomplish this, the fuel tables are set to deliver a constant air/fuel ratio (usually $\lambda = 1$) under a wide range of speeds and loads. With known fuel outputs, it can safely be assumed that any deviation from the desired air/fuel ratio is a result of an error in measured airflow. If the target ratio is $\lambda = 1$, using a wideband air/fuel monitor makes this process easier.

It has been my experience that it is easiest to perform initial fuel corrections in open loop. That way, instantaneous lambda output equals the necessary correction factor needed for the current operation point. The self-correcting action of the OEM lambda controller (closed loop operation) often becomes more of a nuisance than tuning aid. If the stock HEGO is not providing an accurate signal, the PCM's correction may push against the calibrator's initial changes. This can be more prominent in cases where a larger duration camshaft is being used or header lengths have required the HEGO to be moved farther away from the cylinder head.

Most PCMs can be forced to stay in open loop operation by adjusting the closed loop enable tables to values at the extreme of their range. For example, minimum throttle for open loop can be set to 0% or minimum coolant temperature for closed loop set to 300 degrees. After the majority of airflow modeling is complete and fueling errors are inside of a couple percent, changing back to closed loop operation should not significantly change engine operation.

It is possible to perform these corrections in closed loop, but it adds another multiplication step and requires more data recording. In this case, both short- and long-term closed loop fuel trims must also be monitored. These trims must be applied to the actual lambda before calculating the necessary multiplier for airflow corrections. If the closed loop trims have enough authority to bring the engine to $\lambda = 1$, then these corrections themselves become the multiplier to be applied to airflow for correction. This process works best on vehicles where the exhaust manifolds, camshaft, and HEGO installations remain the same as stock. Since the PID (Proportional Integral Derivative) control loop running the HEGO feedback and lambda trim in closed loop is rarely stable at 0% correction, I often find it more useful to simply correct in open loop with a more stable delivered lambda.

The bulk of the work in calibrating the airflow model should be done at nominal temperatures. The engine should be warmed up and any startup enrichment routines should be expired before proceeding. Coolant, oil, cylinder head, and intake air temperatures should be stable to avoid interference from other multipliers while adjusting fuel delivery. Adjustments for cold start, varying air temperatures, and other conditions are handled later in the process. This is simply the time to model the base characteristics of the engine and its airflow measurement.

Mass Air Flow Modeling

For engines that use an MAF sensor, airflow modeling means making sure that the transfer function in the PCM matches the actual output of the sensor. The ideal method is to have data from the exact MAF sensor to be used, as flowed with the exact intake tract from filter to throttle body, and input this data directly into the PCM's MAF table. Since this data is rarely available, the next best thing to do is create this data by actual testing on a dynamometer. Something as simple as a change in air filter design or MAF sensor clocking relative to bends in the plumbing can shift MAF output to the PCM by 20% or more. Verifying actual performance on the exact vehicle in question gives the best accuracy.

Actual MAF output calibration is best done by using the dynamometer to hold the engine exactly at one of the calibration points of the MAF sensor, waiting for the engine to stabilize, and finding the error. The calibration points in question may not necessarily line up with fixed RPM or engine loads. All that is needed is a static airflow rate. This means that simply driving the vehicle in a single gear on a chassis dynamometer can yield a wide range of calibration points simply by changing throttle position and consequently speed. Once the engine stabilizes, the MAF table is adjusted at this point and we move on to the next calibration point. This process is repeated for as much of the MAF sensor range as possible, always waiting for the engine to reach steady state operation before making an adjustment to the MAF table values. If it is not possible to adjust the MAF table in real time, make a spreadsheet showing MAF calibration points, target λ, actual λ, and percent error. (**Figure 9-1**) This makes for fast adjustments to the data tables and a quick return to the dynamometer to confirm or adjust again until error is acceptable.

A word of caution here is to monitor engine temperature and actual lambda. Stop testing if the engine is operating at either high temperature or an excessively rich or lean condition. Running the engine excessively rich ($\lambda > 0.8$) for too long during this process can cause bore wash and ring damage. Likewise, running too lean can generate high exhaust gas temperatures that may damage valves, mani-

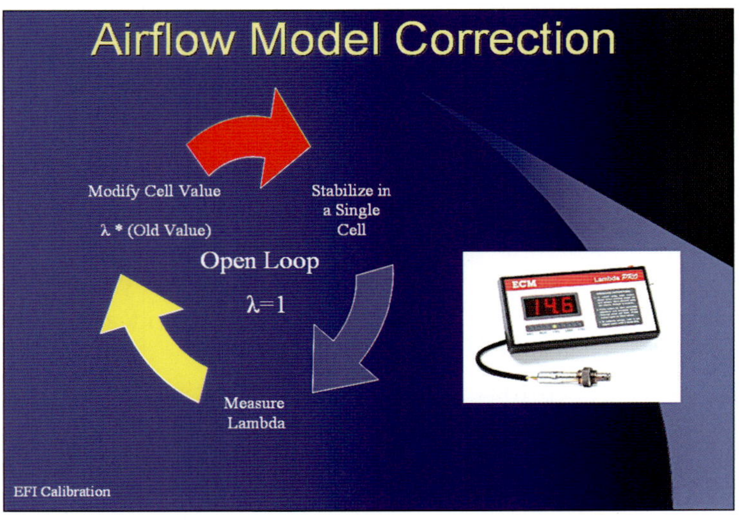

Correcting the airflow model is best done in open loop with a target ratio of lambda = 1 and real time feedback from a wideband. This makes the math for finding the correction factor very simple.

folds, or catalysts. This part of the tuning process is not intended to cover airflow ranges approaching WOT. It should be possible to map the majority of airflow conditions without exceeding about 60% engine load. This should keep temperatures safe.

This is the very coarse tuning part of the calibration process where changes often resemble a lumberjack's chainsaw as opposed to the whittler's knife and sandpaper to be used later. Don't be afraid to stop and make a global 30% or greater change to the MAF curve if it looks like the first guess was really that far off. Look for trends in MAF error to save time. If MAF output appears to be 10% low at all ranges tested so far, go ahead and add 10% all the way up to the maximum values. This saves time later when calibrating the WOT airflow errors.

The process of refining the accuracy of the calculated air mass to match the actual consumption of the engine is repeated until the errors become very small. Less than 5% error is desirable, with less than 3% being ideal. OEM applications usually have less than 1% error across the operating range. The more accurately the airflow modeling is done at this stage, the less work there is to be done later, and the better the vehicle drives. Good calibrators spend most of their time on this part of the process to avoid needing "patchwork" fixes to cover strange engine behaviors later.

MAF Scaling

Some PCMs are coded with a maximum calculated value for total airflow that cannot be changed in the calibrated values. When modifying the engine to make more power,

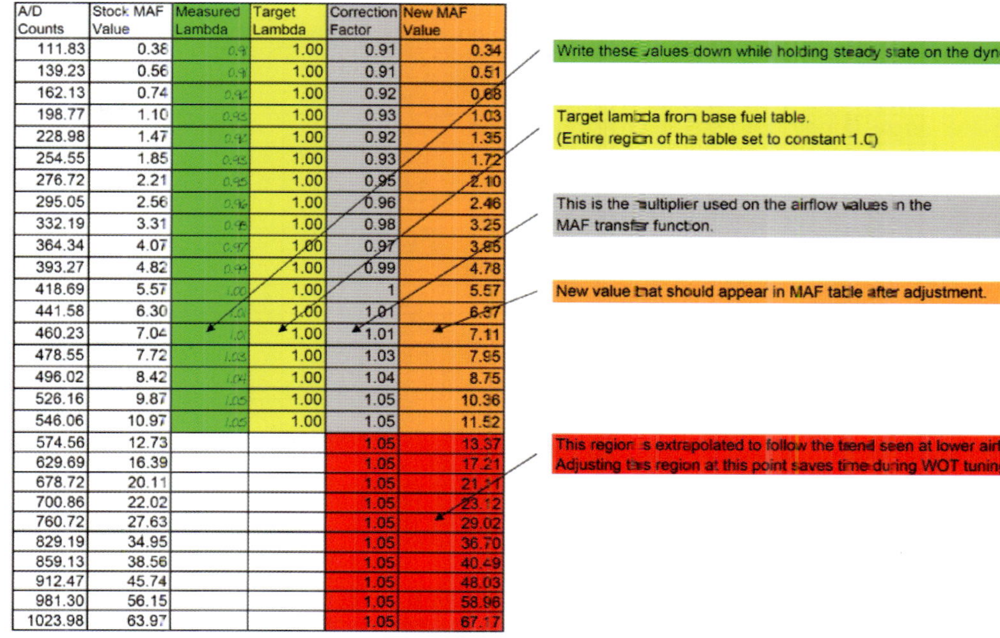

A/D Counts	Stock MAF Value	Measured Lambda	Target Lambda	Correction Factor	New MAF Value
111.83	0.38	0.9	1.00	0.91	0.34
139.23	0.56	0.9	1.00	0.91	0.51
162.13	0.74	0.92	1.00	0.92	0.68
198.77	1.10	0.93	1.00	0.93	1.03
228.98	1.47	0.94	1.00	0.92	1.35
254.55	1.85	0.93	1.00	0.93	1.72
276.72	2.21	0.95	1.00	0.95	2.10
295.05	2.56	0.96	1.00	0.96	2.46
332.19	3.31	0.98	1.00	0.98	3.25
364.34	4.07	0.97	1.00	0.97	3.95
393.27	4.82	0.99	1.00	0.99	4.78
418.69	5.57	1.00	1.00	1	5.57
441.58	6.30	1.01	1.00	1.01	6.37
460.23	7.04	1.01	1.00	1.01	7.11
478.55	7.72	1.03	1.00	1.03	7.95
496.02	8.42	1.04	1.00	1.04	8.75
526.16	9.87	1.05	1.00	1.05	10.36
546.06	10.97	1.05	1.00	1.05	11.52
574.56	12.73			1.05	13.37
629.69	16.39			1.05	17.21
678.72	20.11			1.05	21.11
700.86	22.02			1.05	23.12
760.72	27.63			1.05	29.02
829.19	34.95			1.05	36.70
859.13	38.56			1.05	40.49
912.47	45.74			1.05	48.03
981.30	56.15			1.05	58.96
1023.98	63.97			1.05	67.17

Figure 9-1 *Making a table of MAF breakpoints and the necessary corrections at steady state can ease the tuning process. Any trends can be extrapolated upward to save time later during WOT calibration.*

such as adding a supercharger or larger displacement, this number can be exceeded. In order to retain full function of the PCM, one must find a way to stay within its calculated limits while still moving large amounts of air. The answer is scaling.

If a PCM is limited to 62 lbs/min of airflow (as seen in many GM LS1 applications) and the engine is capable of moving 70 lbs/min of air (not unusual for very high output supercharger systems), some creativity must be employed. The first and most important step is to use an MAF sensor with an accurate range to at least the maximum predicted flow rate. In our example, an MAF sensor with accuracy up to approximately 72 lbs/min is desirable to leave room for a little extra airflow under extreme conditions. To make this new transfer function work in the limited PCM, it is scaled down. In this case, a multiplier of 0.861 (13.9% reduction) is applied to the actual airflow values in the MAF transfer function. That way, even at the new predicted maximum of 72 lbs/min of airflow, the PCM is still within its allowable range.

To compensate for the reduction in calculated airflow, a similar reduction must be made to fuel delivery. Injector size should be reduced by the same amount, a 0.861 multiplier in this case. Do not change the voltage compensation or dead time at this point, since these should not be affected by the scaling.

Changing both the airflow and fuel delivery values keeps the delivered air/fuel ratio on target. However, this scaling also has an effect on load calculations. Even though the PCM is delivering the proper amount of fuel to go with the actual air entering the engine, the calculated load is not correct at this point. Load must be corrected since many of the base fuel and spark tables use calculated load as a primary input. To correct the load calculations, the

Trim / MAP																

Scaled Engine Speed RPM

Trim / MAP	200	700	1000	1200	1800	2200	2600	3000	3400	3800	4200	4600	5000	5400	5800	6200
13.44	0.230	0.240	0.240	0.206	0.221	0.255	0.270	0.284	0.284	0.304	0.314	0.304	0.304	0.289	0.284	0.284
19.29	0.240	0.240	0.309	0.245	0.260	0.275	0.299	0.319	0.348	0.353	0.358	0.373	0.373	0.368	0.368	0.338
25.14	0.255	0.255	0.270	0.270	0.309	0.333	0.368	0.377	0.397	0.402	0.412	0.417	0.412	0.422	0.422	0.397
31.00	0.265	0.275	0.284	0.333	0.363	0.392	0.387	0.431	0.436	0.446	0.451	0.456	0.456	0.456	0.456	0.436
36.85	0.343	0.368	0.373	0.407	0.441	0.446	0.475	0.510	0.529	0.539	0.539	0.549	0.549	0.549	0.549	0.529
42.70	0.382	0.407	0.480	0.515	0.539	0.539	0.544	0.578	0.588	0.593	0.588	0.588	0.588	0.559	0.574	0.525
48.55	0.368	0.407	0.495	0.578	0.569	0.569	0.574	0.598	0.627	0.632	0.662	0.672	0.662	0.632	0.642	0.569
54.40	0.397	0.402	0.529	0.608	0.637	0.637	0.632	0.632	0.667	0.672	0.706	0.716	0.706	0.691	0.662	0.588
60.26	0.412	0.456	0.564	0.618	0.642	0.642	0.642	0.652	0.696	0.701	0.745	0.755	0.750	0.730	0.711	0.662
66.11	0.461	0.461	0.564	0.642	0.652	0.652	0.652	0.662	0.725	0.730	0.784	0.770	0.765	0.750	0.725	0.701
71.96	0.471	0.471	0.583	0.642	0.686	0.686	0.686	0.696	0.750	0.765	0.799	0.765	0.760	0.755	0.706	0.681
77.81	0.510	0.510	0.588	0.657	0.686	0.686	0.686	0.711	0.765	0.770	0.809	0.775	0.770	0.760	0.696	0.676
83.66	0.534	0.534	0.608	0.686	0.691	0.691	0.701	0.725	0.779	0.784	0.814	0.779	0.775	0.750	0.696	0.657
89.52	0.559	0.549	0.574	0.642	0.716	0.721	0.721	0.735	0.789	0.794	0.814	0.824	0.799	0.799	0.770	0.706
95.37	0.603	0.564	0.657	0.716	0.760	0.770	0.755	0.770	0.824	0.833	0.858	0.843	0.814	0.799	0.789	0.730
101.22	0.603	0.613	0.642	0.765	0.775	0.779	0.779	0.853	0.868	0.868	0.868	0.848	0.819	0.833	0.789	0.735

Manifold Pressure KPa

Scaled Eng. Spd	MAP	Base Vol. Effcy.	Target A:F	Knock Retard	Duty Cycle
0	13.4	0.000	10.00	0.00	0.0

PW Est. | AutoCal | Overlay | 3D Graph

[ ] Change selected cell values +,P Add to selected cells *,X Multiply selected cells	\ Increase/Decrease rate of change SHIFT+Arrows Select cells T Toggle Trace Mode / U Clear Trace	ENTER Send changes to ECM SPACE Select current operating point CTRL-Z Undo last operation

RPM	SPARK	MAP	ECT	HEGO	O2 FBK	Closed Loop Correction	Error Code	Offline (F9)	
0	0.00	0.00	0	0.00	0.0	Rich … Lean	99		HELP (F1)

-25 -20 -15 -10 -5 0 5 10 15 20 25   Clear (F8)   Default

The basic component of the speed density airpath model is the VE table. This Accel DFI GenVII table shows relative VE values versus speed and MAP.

calibrated engine displacement must also be reduced by the same amount.

With all of this done, the PCM now sees an exact scale model of everything. It takes what it thinks is 86.1% of actual airflow and delivers a theoretical 86.1% of matching fuel flow to an engine 86.1% of the actual size yielding exact desired lambda and load calculations. The primary benefit to this scaling technique is that any future fueling corrections can still be performed on a percentage basis. This means that even on a scaled MAF calibration, a 10.0% change to desired lambda still yields a 10.0% change to delivered lambda as long as adjustments are made as multipliers. Although a slight reduction in airflow measurement precision (13.9% in this example) results from this method, OEM transfer functions usually have more than

enough break points to accurately model the sensor input. The added accurate measurements at high flow rates outweigh the minor drawback of precision loss across the spectrum.

This method is so effective that it has even been used by Ford in the OEM calibration of the high output GT supercar using a standard EEC-V processor. Keep in mind that this is a vehicle making over 550 hp with 91-octane fuel while passing all CARB and Federal emissions standards.

Speed Density Airflow Modeling

For speed density systems, airflow modeling means building the PCM's volumetric efficiency reference table to accurately represent actual mass flow at all speed and load points of the engine. Load is

expressed in terms of manifold absolute pressure, received directly from the MAP sensor. Absolute pressure gives the PCM the density of the air mixture in the manifold, so mass flow can be calculated using the basic gas law once temperature is known. Some systems use Alpha-N mapping that replaces MAP load values with TPS position.

Much of the process is the same as used in building the MAF table, except that the dynamometer is used to hold the engine at various speed and load intersections rather than constant airflow points. To do this, speed is usually set at one of the calibration points in the base fuel map first. After the engine stabilizes at the lowest load point for that speed, changes are made to the corresponding load point. Speed is held constant, throttle is opened to increase load to the next point, and the process is repeated until engine or exhaust temperatures prohibit. All of this mapping should be done at a constant inlet air temperature. Changes in IAT significantly affect the density of the air being ingested by the engine, so it is best to attempt to keep IAT steady during VE mapping.

Much like mapping the MAF sensor transfer function, it is helpful to recognize trends in the volumetric efficiency table. This table should not really have any drastic changes between cells. When viewed as a three dimensional map, it should appear generally smooth with higher values near the engine's torque peak, where efficiency is highest.

One key that is often overlooked by aftermarket calibrators is resolution of speed density maps. If the PCM has a limited number of cells in the volumetric efficiency table, it is important to choose where these points are best served in the calibrator. The VE table represents curves of engine efficiency versus speed and load. While some engines can be mapped with evenly spaced calibration points every 500 rpm or so, many cannot. To properly map these curves, it is necessary to plot enough points near the tightest bends to reduce errors. In areas where engine efficiency can change rapidly, such as just off idle, points should be set closer to each other in order to properly represent actual airflow changes. This gives the PCM better resolution to compensate for changing efficiency as the intake manifold and camshaft tuning becomes more influential as speed increases. I have often found it useful to have set points for speed at crank: 100 rpm below target idle speed, idle speed, 100 rpm above idle speed, and a few more added below 3,000 rpm to improve resolution. The changes become more linear at higher engine speeds, so it doesn't hurt accuracy very much to map this region with points that are slightly farther apart.

If the camshaft being used only allows the engine to develop a limited amount of vacuum, the load scale should also be adjusted to reflect the actual range of engine inputs. There is no benefit to having a VE table with 10, 20, 30, and 40 kPa load points if the maximum vacuum seen is only 55 kPa due to cam design. In this case only one load point for extreme vacuum should be necessary, leaving room for more resolution between 50 kPa and 100 kPa. Better planning for set point distribution here can help alleviate drivability problems later when the PCM has trouble interpolating between two widely spaced points. In this case, a middle point to better define the curve would help. Most engines exhibit quasi-linear efficiency changes with load, so spacing can be even on the load scale as long as the range is reasonable.

When first building a base VE table from scratch, the calibrator can safely assume that values increase with higher load and RPM. This means that if part load testing yields a VE of 60% at 70 kPa and 3,000 rpm, that same 60% value can be copied across to the remaining RPM cells at 70 kPa for an initial guess. This gets the calibrator a closer guess before attempting to fine-tune VE at 3,500, 4,000, 4,500, etc. The trend should continue upward as well. Taking our example of 60% VE at 70 kPa at 3,000 rpm, we also know that VE values should steadily increase from 60% as we work upward in load at 3,000 rpm. It may be possible to perform WOT (≈ 100 kPa on a naturally aspirated engine) calibrations at low engine speeds in this manner. But as engine speeds get higher, temperatures climb quickly at high load, preventing steady state mapping of those cells. Follow the trend to develop a best guess at what the values might be near 100 kPa at higher engine speeds to save time during WOT calibration later.

After the base volumetric efficiency map has been constructed under stable conditions, the vehicle can be checked at different inlet temperatures to build the IAT compensation curve. If tuning was performed with steady 30 degree F inlet temperatures on a dynamometer and the vehicle exhibits a lean condition when driven outside in 50 degree F ambient conditions, the calibrator needs only to adjust the IAT fuel multiplier to compensate. Changing the base VE table at this point only causes a rich condition when inlet temperatures rise again. To further aid the process, the vehicle should later be allowed to completely cool (preferably overnight) to check the bottom end of this compensation curve for accuracy. If it is not possible to actually test with very cold inlet temperatures, a straight-line extrapolation usually works.

Spark Advance

After the majority of the airflow modeling has been completed, the calibrator can now look into spark mapping. For part load conditions, it is usually best to find MBT timing for most cases. This allows the engine to operate as efficiently as possible with optimal fuel economy and good throttle response. The spark map

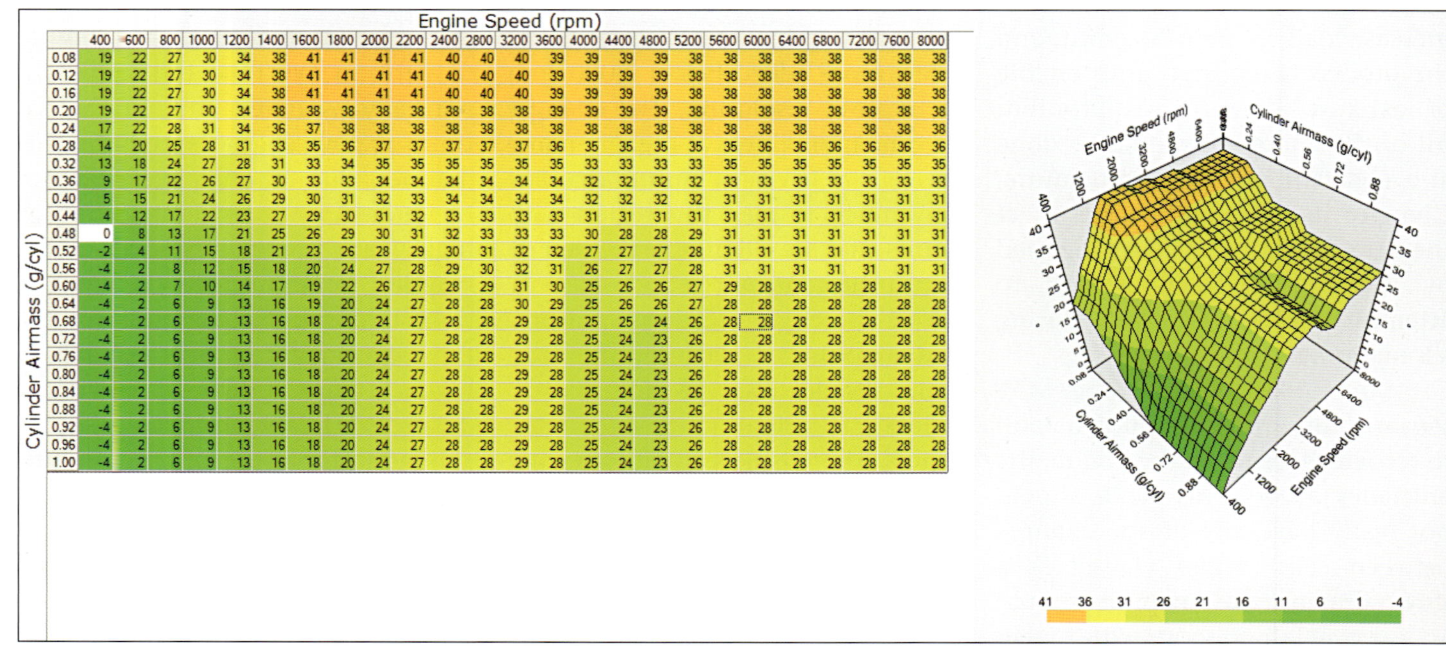

Cylinder Airmass (g/cyl)	400	600	800	1000	1200	1400	1600	1800	2000	2200	2400	2800	3200	3600	4000	4400	4800	5200	5600	6000	6400	6800	7200	7600	8000
0.08	19	22	27	30	34	38	41	41	41	41	40	40	40	39	39	39	39	38	38	38	38	38	38	38	38
0.12	19	22	27	30	34	38	41	41	41	41	40	40	40	39	39	39	39	38	38	38	38	38	38	38	38
0.16	19	22	27	30	34	38	41	41	41	41	40	40	40	39	39	39	39	38	38	38	38	38	38	38	38
0.20	19	22	27	30	34	38	38	38	38	38	38	38	38	39	39	39	39	38	38	38	38	38	38	38	38
0.24	17	22	28	31	34	36	37	38	38	38	38	38	38	38	38	38	38	38	38	38	38	38	38	38	38
0.28	14	20	25	28	31	33	35	36	37	37	37	37	37	36	35	35	35	35	35	36	36	36	36	36	36
0.32	13	18	24	27	28	31	33	34	35	35	35	35	35	35	33	33	33	33	35	35	35	35	35	35	35
0.36	9	17	22	26	27	30	33	33	34	34	34	34	34	34	32	32	32	32	33	33	33	33	33	33	33
0.40	5	15	21	24	26	29	30	31	32	33	34	34	34	34	32	32	32	32	31	31	31	31	31	31	31
0.44	4	12	17	22	23	27	29	30	31	32	33	33	33	33	31	31	31	31	31	31	31	31	31	31	31
0.48	0	8	13	17	21	25	26	29	30	31	32	33	33	33	30	28	28	29	31	31	31	31	31	31	31
0.52	-2	4	11	15	18	21	23	26	28	29	30	31	32	32	27	27	27	28	31	31	31	31	31	31	31
0.56	-4	2	8	12	15	18	20	24	27	28	29	30	32	31	26	27	27	28	31	31	31	31	31	31	31
0.60	-4	2	7	10	14	17	19	22	26	27	28	29	31	30	25	26	26	27	29	28	28	28	28	28	28
0.64	-4	2	6	9	13	16	19	20	24	27	28	28	30	29	25	26	26	27	28	28	28	28	28	28	28
0.68	-4	2	6	9	13	16	18	20	24	27	28	28	29	28	25	25	24	26	28	28	28	28	28	28	28
0.72	-4	2	6	9	13	16	18	20	24	27	28	28	29	28	25	24	23	26	28	28	28	28	28	28	28
0.76	-4	2	6	9	13	16	18	20	24	27	28	28	29	28	25	24	23	26	28	28	28	28	28	28	28
0.80	-4	2	6	9	13	16	18	20	24	27	28	28	29	28	25	24	23	26	28	28	28	28	28	28	28
0.84	-4	2	6	9	13	16	18	20	24	27	28	28	29	28	25	24	23	26	28	28	28	28	28	28	28
0.88	-4	2	6	9	13	16	18	20	24	27	28	28	29	28	25	24	23	26	28	28	28	28	28	28	28
0.92	-4	2	6	9	13	16	18	20	24	27	28	28	29	28	25	24	23	26	28	28	28	28	28	28	28
0.96	-4	2	6	9	13	16	18	20	24	27	28	28	29	28	25	24	23	26	28	28	28	28	28	28	28
1.00	-4	2	6	9	13	16	18	20	24	27	28	28	29	28	25	24	23	26	28	28	28	28	28	28	28

Most PCMs use speed and load for inputs to the base ignition map. The actual delivered timing may vary from this if other adders for temperature, lambda, or torque control are active.

should look like an inverse of the volumetric efficiency map, smooth and predictable. Using the dynamometer to find best torque output at a few points along the load and RPM curves, as well as filling in the remaining cells of the base spark table, is rather simple.

If spark advance is left too low in part throttle regions, the result can be a bucking or "trailer-hitching" sensation on the road as the engine misfires. Remember that larger camshafts increase the natural EGR of the engine and require more spark advance to compensate for the cooler combustion temperatures. High-octane fuels also require a global increase in ignition advance to compensate for their slower burn rate.

It is not usually necessary to plot every single load and speed point to find MBT unless working on an actual OEM release calibration. For areas outside the easily measurable range, extrapolate the trends seen in the known areas of the map. Timing should trend downward with increasing load and upward with engine speed.

During this process, monitor the IAT and ECT to ensure that the engine is operating near the nominal values. Actual temperature compensation curves can be finished after the base tables are complete. Don't be afraid to use what looks like large values for ignition advance at part load—this is often what the engine wants. Running as much spark advance as possible (up to MBT) results in higher overall engine efficiency, more torque, and crisper throttle response.

Knock is still a concern at part load. Although its presence is less likely at lower loads, it should still be avoided. Most engines exhibit a slight loss in torque at part load before actually hitting the knock limit. If steady

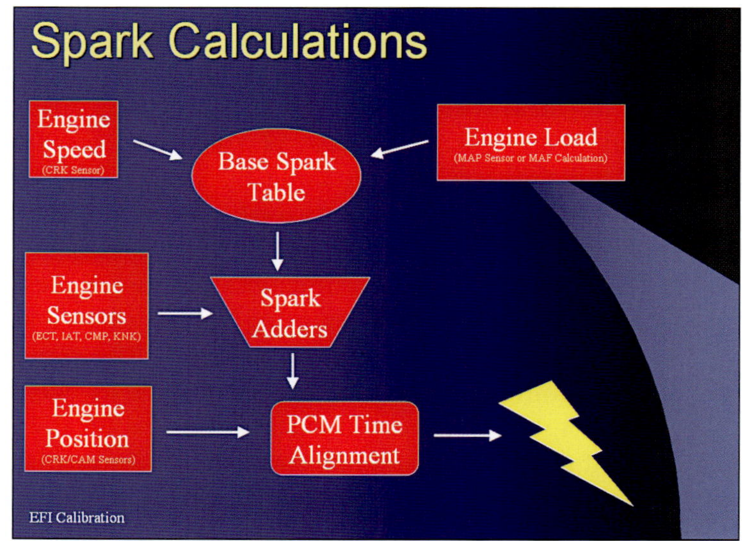

The PCM calculates total timing primarily from speed and load. Additional adjustments are made to correct for temperatures, fueling, and other conditions before aligning the spark event with the crank sensor output.

state testing is being performed on a dynamometer, this can be seen while adjusting the timing in the search for MBT. Listen carefully for any signs of knock and reduce timing accordingly in the affected area.

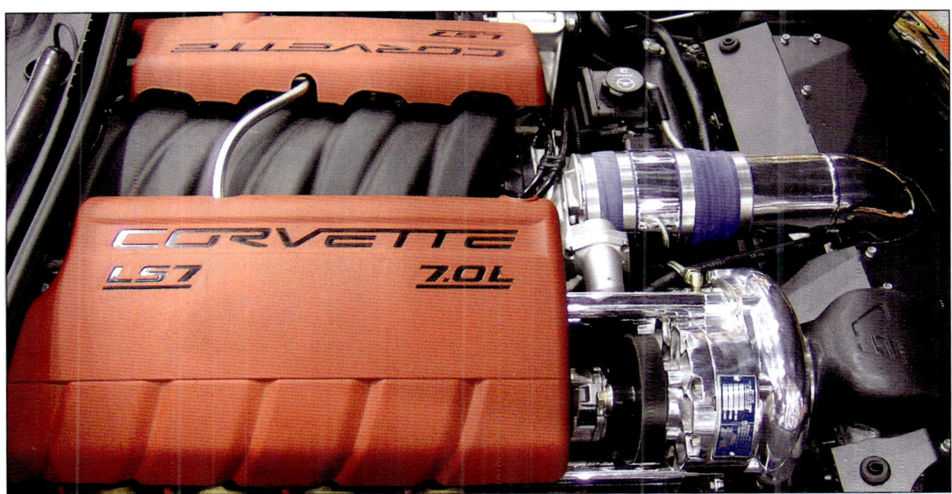

High compression and superchargers are seldom friends. This kit from Vortech engineering for the 11.0:1 Corvette Z06 works very well as a result of low charge temperatures and careful engine tuning.

Chassis Dyno Use

Modeling airflow is the most important task in engine calibration. Creating this airflow model is much easier if one is not trying to hit a moving target. The ability to hold the engine steady in one operating state means that different points in the airflow model can be isolated, measured, and tuned individually. A load bearing chassis dynamometer gives the calibrator the tool needed to perform these tasks.

For mass air systems, the dynamometer simplifies MAF transfer function mapping by allowing the operator to hold a constant airflow rate. This is done by running the vehicle at cruise and maintaining a consistent speed in one gear. If speed is held constant with a consistent load from the dynamometer, actual airflow through the engine will stabilize. The datalogging tool should display a consistent value for actual mass airflow rate at this point. It is necessary to hold a constant speed and load to minimize oscillations in fuel delivery from wall film or acceleration-enrichment functions. This may take a few moments, so watch the display of the wideband and wait for a stable reading. If a stable reading cannot be done, it may be wise to choose another speed/load point for that airflow rate. Changing to a different gear on the chassis dyno will usually do the trick.

Actual output from the MAF sensor is noted along with delivered lambda as seen on a wideband meter. If the injector flow characteristics have already been properly entered, any resulting error between actual and target lambda at this point is likely from the inaccuracy of the MAF transfer function. Holding the engine at one point along this transfer function allows the calibrator to correct the airflow values (in the MAF transfer function) to closely match actual airflow through the engine for each of these points. Corrections to the actual airflow values in the MAF transfer should be done using the following equation:

$$\text{Correction Factor} = \frac{\text{(Actual Lambda)}}{\text{(Target Lambda)}}$$

The new airflow rate to be entered into the MAF transfer function is determined by:

$$\text{New Airflow Value} = \text{(Correction Factor)} \times \text{(Current Airflow Value)}$$

If target lambda is set to 1.00, the math is significantly easier. The correction factor equals Actual Lambda when Target Lambda is equal to 1.00, so we have:

$$\text{New Airflow Value} = \text{(Actual Lambda)} \times \text{(Current Airflow Value)}$$

When the MAF transfer function matches the actual airflow rate of the engine, delivered lambda equals the target value. The lower half of the MAF transfer function can be mapped in this manner, usually with a target lambda value of 1.00 just to make the math easier. The lower half of the MAF transfer function covers about 90% of driving

Continued

Chassis Dyno Use Continued

conditions for the vehicle, so more precision here leads to smoother overall operation on the street or track.

The top end of the MAF transfer function is usually found by performing a WOT pull under load. The WOT pull is done in the conventional manner on the dyno, but special care is taken to ensure proper recording of the MAF output data versus RPM. At the end of the measurement, delivered lambda values as recorded by the wideband and dynamometer are compared with the target values in the PCM to find any error between the two. When the target values for WOT lambda are constant, the math for transfer function corrections is the same as before. The primary difference is that the target value is most likely not going to be $\lambda = 1$, so the correction factor needs to be calculated from delivered lambda.

For speed density systems, the load bearing dynamometer is a tremendous time saving tool. The primary inputs to the PCM's airflow model are engine speed and load. The load bearing dynamometer is designed to control these two variables very precisely. During base calibration exercises on speed density systems, the tuner wants to hold a specific speed load point to check the accuracy of the airflow model. This is done by commanding a "constant speed test" on the dyno. The dyno is set to hold a consistent speed, regardless of input power from the engine. This allows the calibrator to adjust engine load (the other input to the base fuel map) simply by changing throttle position.

The result is a condition where the dyno does the work to hold a steady speed and the calibrator's right foot holds a steady load. Once isolated to a single cell in the volumetric efficiency table, work can resume on correction of the airflow model. Again, it is important to reach a stable condition of both calculated airflow and fuel delivery to avoid confusing base VE table adjustments with wall film or AE input.

At this point, actual lambda is checked with a wideband meter and compared against the target value in the PCM. Corrections are made to the values in the VE table until the delivered lambda is equal to the values in the target lambda table. The calculations are the same for VE value corrections as they were for MAF transfer value corrections. This process is repeated for all cells in the VE table by changing load with throttle or speed with the dynamometer set point. If the VE table corrections are performed with enough diligence, there will be a minimum of work to do later when attempting to smooth out drivability concerns.

Temperature almost certainly climbs too quickly to allow for steady state WOT tuning on the chassis dynamometer. To properly adjust these high load cells of the VE table, a WOT sweep is run in a similar fashion to MAF tuning. Again, careful attention is paid to recording the load input versus RPM in order to identify which cells were used in the fuel calculation during the run. Delivered lambda is compared to the target lambda and a correction factor is calculated in the same manner.

Whether a speed density or mass air system is employed, the same general procedure is used to find the ideal ignition set points at part load. Speed is held constant and a specific load condition is "dialed in" using throttle input. The dynamometer should be configured to show instantaneous torque output clearly on the display. This value is noted and ignition advance is adjusted. The ignition advance can be increased up to the point where either torque is maximized (this, by definition, is MBT timing) or knock is detected. At light loads, this is the ideal spark advance that allows the engine to run most efficiently. The closer the engine runs to MBT timing, the better fuel economy can also be expected at light loads.

Keep in mind that light loads generate correspondingly low cylinder pressures. In order to make peak torque at light loads, the actual spark advance values for these regions may seem quite large compared to "total timing" at WOT. Increases in speed only exaggerate this, so don't be afraid of what appear to be large ignition advance values at high speed and low load. Just remember to check actual engine torque output and respect the knock limit.

Wide-open throttle ignition tuning on the dyno is usually done in sweeps much like lambda tuning. Again, temperature limits typically prevent steady state testing. Verification of proper ignition advance at WOT can be done by comparing torque curves from two different sweep tests of the same engine. Since torque is the best indicator of cylinder pressure, spark advance changes that had a positive effect on BMEP show up clearly on a torque trace. Provided no knock was present on either run, selecting the proper ignition advance value for each speed break point is as simple as seeing which torque trace was higher and keeping the appropriate value in the ignition table.

SETTLING DOWN

Now that the engine has been mapped under most stable operating conditions, it is possible to move on to a less stable condition: idle. Because of the slow engine speed, there is a relatively long time between possible corrections the PCM can make at each TDC event. The longer pause before the next feedback signal makes it easy to over- or under-correct at idle speeds. Following trends found under stable operating conditions at medium engine speeds and loads downward gives a better estimate of what the engine wants at idle speed. Remembering that idle speed is the result of the delicate balance between engine torque and engine drag from loads, small changes in engine torque can yield relatively large changes in idle speed.

A couple layers of control are required for this delicate balance at the lowest acceptable speed without stalling. The largest controlling factor to idle speed is engine throttling. Throttling can come from either the throttle blade itself or the IAC motor. While very large changes can be made to engine torque by changing throttling, actual changes take a relatively long time to fill the manifold and allow the engine to react.

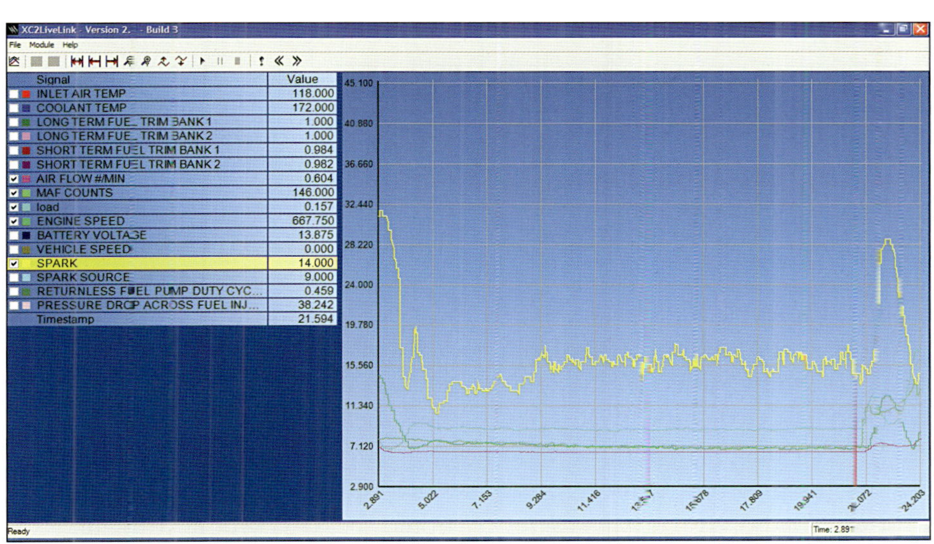

A datalog of engine behavior at idle shows relatively low calculated engine loads and airflows. Notice how spark advance moves rapidly to help maintain a steady engine speed.

To prevent stalling in the meantime, a faster correction to engine torque can be made by changing spark advance. This is why it is desirable to set base ignition timing lower than MBT at idle, leaving room to adjust timing when necessary to reduce speed fluctuations and prevent stalling. Idle spark advance values can be surprisingly small, especially in engines with more efficient combustion chamber designs. Many modern DOHC engines with stock cams idle with single digit spark advance.

If a larger than stock camshaft design simply does not allow stable idle at the stock speed, adding 100 rpm at a time can quickly bring it into stable operation. Once stable, reducing an additional 50 rpm may help quiet engine operation at idle. This is where carefully choosing load and speed set points in a speed density application or MAF scaling come in very handy. There is no reason to simply increase

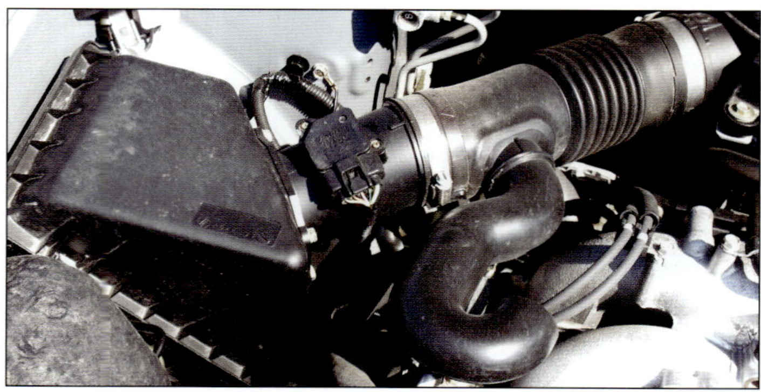

The MAF on this engine is installed on the airbox. The hook-shaped appendage on the inlet tube is a Helmholtz resonator, used to cancel out standing wave vibrations that would otherwise corrupt the MAF signal at low speeds as well as make an audible noise.

This engine uses two Helmholtz resonators (arrows) to further dampen intake tract oscillations. Each chamber is tuned to a different frequency. The PCM is hiding beneath the right arrow.

idle speed from 800 rpm to 1,000 rpm after a camshaft change when a little extra calibration time allows the engine to be stable at 850 rpm. Careful control of fuel delivery, airflow mapping, and spark compensation are the keys.

Little by little, engine speed can be reduced to continue mapping airflow from the stable areas mapped earlier to speeds approaching desired idle. As long as the fuel injector properties have been modeled correctly, it is possible to continue with airflow mapping for progressively slower speeds and lower loads. With each progressively lower engine speed, the airflow model can be adjusted as before. The same process of checking actual lambda against the target value applies. With proper airflow modeling, the engine is likely to exhibit loads at idle of 10 to 18%.

Changes in injector offset (dead time) show up most prominently at low speed where actual pulsewidths are smallest and dead time represents a larger percentage of total injection time. If the injector flow was not modeled accurately before starting, this is where the calibrator sees issues.

When starting from scratch on a throttle stop set point without ETC, allow the engine to idle with the blade held slightly open. Remove any IAC control (unplug the motor from the harness) and slowly close the blade position until the engine barely runs near the desired speed. Reinstalling the IAC harness lead should result in a more stable idle. This prevents the IAC from leaning to either a fully open or closed static position. The object is to get the IAC motor in the middle of its adjustment range so that it has maximum flexibility to adjust airflow either up or down. The TPS output should be checked at this time to ensure that the output value is within normal operating range and registers as "closed throttle" to the PCM. In some cases, it is not possible to open the throttle blade enough and remain within the range the PCM considers "closed throttle." In these rare instances, a small bleed hole can be drilled in the throttle blade to allow slightly more air bypass when closed. Start small and increase in very small increments to avoid scrapping the throttle body. A final check should be made to verify that commanded idle speed in the PCM matches actual engine speed and that the IAC is near the middle of its range.

During idle speed calibration, it is helpful to monitor data in real time for MAF, MAP, lambda, injector pulsewidth, and temperatures. Verify that all signals are within normal operating range for the PCM. Some PCMs have a value for minimum allowed MAF that may need to be reduced when using a scaled meter. Actual MAF or MAP values give an indication of exactly which cells should be modified to complete airflow modeling at idle speeds. Pulsewidth is also useful to monitor since it can show any "hunting" activity in the PCM for target lambda that may be caused by background functions such as evaporative or EGR compensations. Some PCMs also have a lower limit value for pulsewidth that may need to be reduced when using larger fuel injectors.

After finding the stable idle speed at normal operating temperature, a temperature compensation curve can be built. To offset the poor combustion, higher fluid viscosities, and undesirable emissions at cold start, idle speed should be increased at low engine temperatures. If idle is stable at 800 rpm and 190 degrees F, this idle speed is usually fine down to about 160 degrees F. Expect to add up to 400 rpm at freezing to allow for smooth engine operation, with a smooth transition as temperature changes. A side benefit to increased engine speed at cold temperatures is faster warm-up times for components, coolant, and oil. As a measure of precaution, temperatures above 220 degrees F may also be aided by increasing idle speed 100 rpm or so to boost water pump flow during cool-down after extreme operation.

Dashpot

After a stable idle has been found, the PCM needs a way to get to it smoothly. Just like a carburetor, most PCMs have a dashpot function built in that allows for a softer landing to the actual idle speed. Dashpot is usually shown as bypass airflow from the IAC at closed throttle versus engine speed. It is up to the calibrator to contour this function so the engine speed drops quickly once the throttle is closed at higher RPM, and slows its rate of change as idle is approached. Too fast and the engine drops right past idle speed and stalls; too slow and the engine hangs at higher RPM when lifting off the throttle. Changes in throttle body size, camshaft design, and intake manifold design and volume have noticeable effects on required dashpot.

Lower dashpot values at medium to high engine speeds can be used to add engine braking when the throttle is lifted. Care should be taken to make sure this does not interfere with normal cruise throttle position, which may be relatively low for larger throttle bodies.

When adding a supercharger to a naturally aspirated engine, it is important to recognize the source of air feeding the IAC mechanism. When the throttle is closed, the intake manifold is under vacuum and higher pressure is

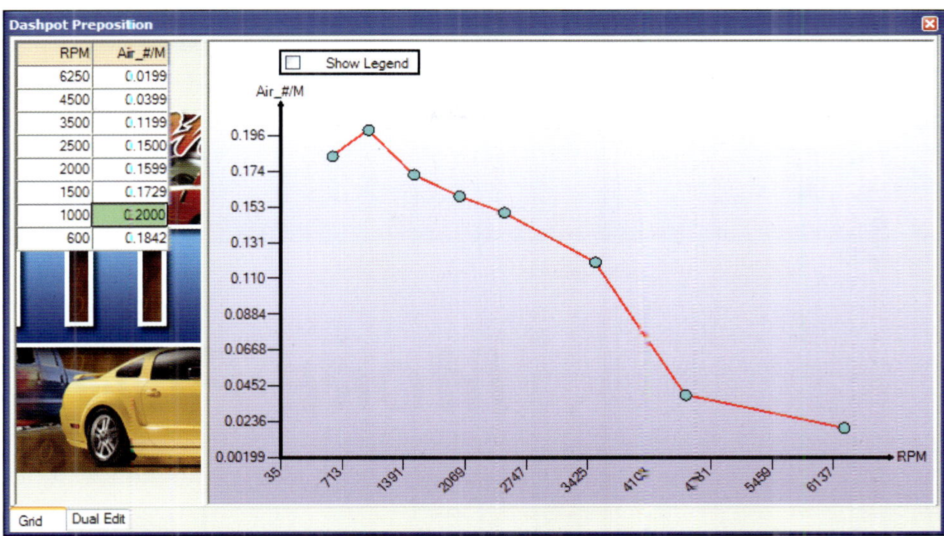

RPM	Air_#/M
6250	0.0199
4500	0.0399
3500	0.1199
2500	0.1500
2000	0.1599
1500	0.1729
1000	0.2000
600	0.1842

The dashpot function controls the amount of airflow at closed throttle. At higher speeds more air is allowed through the engine to control pumping losses and drag.

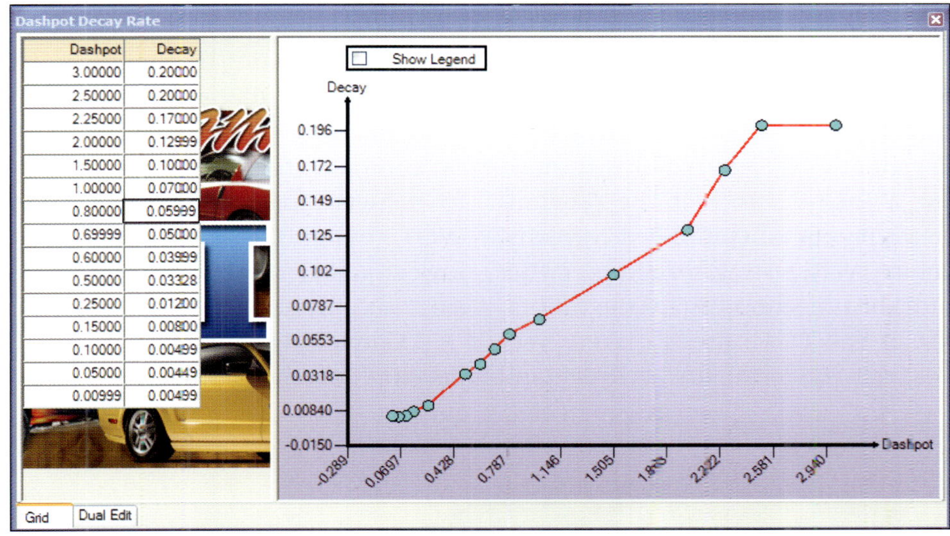

Dashpot	Decay
3.00000	0.20000
2.50000	0.20000
2.25000	0.17000
2.00000	0.12999
1.50000	0.10000
1.00000	0.07000
0.80000	0.05999
0.69999	0.05000
0.60000	0.03999
0.50000	0.03328
0.25000	0.01200
0.15000	0.00800
0.10000	0.00499
0.05000	0.00449
0.00999	0.00499

The actual amount of dashpot airflow must be reduced over time for the engine to return to idle. This table slows the rate of reduction as speeds decrease to provide a gentle return to idle speed without stalling.

present in front of the throttle blade. This pressure differential is exaggerated in supercharged applications. It is desirable to have the IAC draw air from an atmospheric source rather than the pressurized section between compressor and manifold. This reduces the force-feeding effect on the IAC, keeping its actual flow range low where it belongs, and increases the adjustment precision of the PCM.

Dashpot values need to be greatly reduced when the IAC is fed from a pressurized source. The same area of opening in the IAC with higher-pressure differential across it yields greater flow in the exact same manner as increased fuel rail pressure increases injector flow. This also means that each "step" size of adjustment to the IAC position is larger, reducing resolution and control.

MORE POWER!

The next step is the fun part: wide-open throttle. Many tuners jump straight to this part too early. It's easy to understand why. WOT is F-U-N when the engine makes a lot of power, and there is usually some degree of glory associated with making a lot of power. If the previous steps have been completed correctly, actual WOT testing should go rather quickly and smoothly for the calibrator. If the PCM already has a very accurate model of fuel delivery and good representation of actual air mass flow, striking the correct ratio with large flow numbers isn't difficult at all. Spending more time in the early stages of calibration to build an accurate model of airflow can reduce or eliminate the guesswork needed to achieve the desired lambda at WOT. When this procedure is followed, actual WOT calibration takes surprising little time.

Conversely, tuners who fail to develop good mapping of actual engine performance at lower loads often tend to take a long time to find the right combination at WOT. The quicker the proper settings can be found at WOT, the less abuse the engine is subjected to during the testing process. Since engines rarely fail at low load, it is advisable to make an effort to get the high-load tuning right in as few attempts as possible. In short, if you have not already completed part throttle mapping, don't bother trying to tune the WOT maps. The calibrator who skips directly to WOT tuning without mapping part load areas first almost certainly has a much more difficult time blending the maps later for good drivability and transition to power.

The actual WOT tuning process is similar in concept to part load. The fuel delivery remains consistent with the model used at medium and low loads. The airflow model must now be adjusted in the upper range to match actual mass flow through the engine. Again, since the actual fuel delivery is known, the assumption is made that any difference between the commanded lambda and delivered lambda is a result of error in the

WOT testing on a chassis dyno allows the calibrator to collect large amounts of data for careful review after each run. (Nate Tovey)

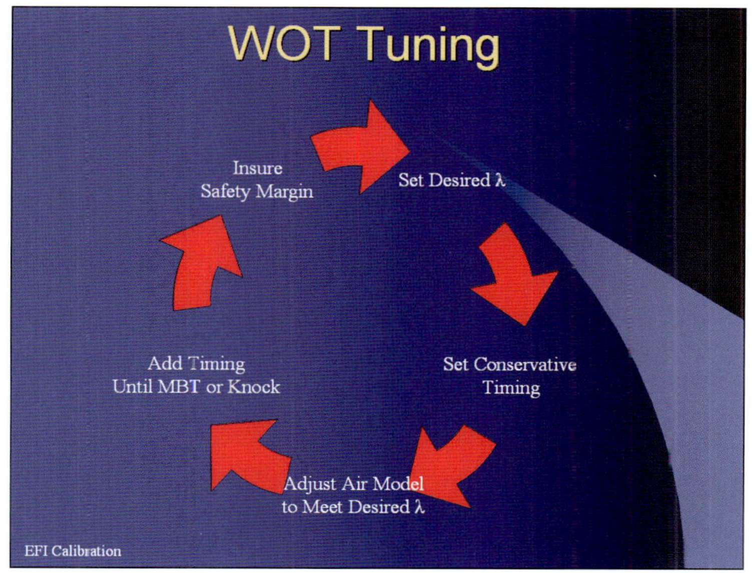

The basic process for WOT tuning is much like part throttle tuning. The primary difference is that the target A/F ratio is usually much richer than stoichiometric. (Nate Tovey)

mal operating temperature and run in a WOT sweep from ~2,000 rpm to redline, provided there are no signs of detonation or lean mixture. During this run, a datalogger is used to record RPM, MAF, MAP, and actual lambda. If the engine achieves $\lambda = 0.85$ at 2,500 rpm, no change is necessary for those cells. If a lean condition of $\lambda = 0.9$ is seen at 4,000 rpm, an adjustment is calculated by dividing actual lambda by target lambda and using this as a multiplier for the corresponding airflow value. If a mass air system is used, this number is multiplied by the value in the MAF transfer function that was recorded at 4,000 rpm. If the value recorded by the datalogger is 3.2 v at 4,000 rpm, the curve is adjusted locally by:

$$\text{Airflow correction} = \frac{(0.9)}{(0.85)} = 1.0588$$

This yields a ~6% correction to add to the airflow number in the MAF transfer function at 3.2 v. Looking at

airflow model. Just like at part load, this error is multiplied by the predicted airflow to bring it in line with actual flow. An accurate, temperature-compensated wideband oxygen sensor is absolutely necessary for accurate measurements during the WOT tuning process. Standard HEGOs lack accuracy in the target lambda range for these tests and only serve to mislead and confuse.

A safe (lower than the expected maximum) amount of ignition timing is chosen and target lambda is set to a slightly rich value. Running slightly rich and with less-than-ideal ignition advance to start reduces the risk of detonation and helps keep exhaust gas temperatures safe. I usually use $\lambda \approx 0.85$ (12.5:1 a/f) for naturally aspirated engines and $\lambda \approx 0.77$ (11.3:1 a/f) for supercharged engines to start. It is helpful to change large areas of the load maps for desired lambda to the target WOT ratio at first. That way, even if calculated load at WOT is relatively low at low RPM, the target ratio remains the same. This simplifies the math when correcting MAF or VE values to achieve the desired ratio. Other fuel adders

such as catalyst and piston protection strategies should be disabled for this part of the testing. They can be reactivated after the completion of WOT airflow mapping if needed.

For example, a naturally aspirated engine is desired to be run at $\lambda = 0.85$ under WOT conditions. First, all target lambda values in the fuel map are set to 0.85 for loads over 50%. The engine is warmed to nor-

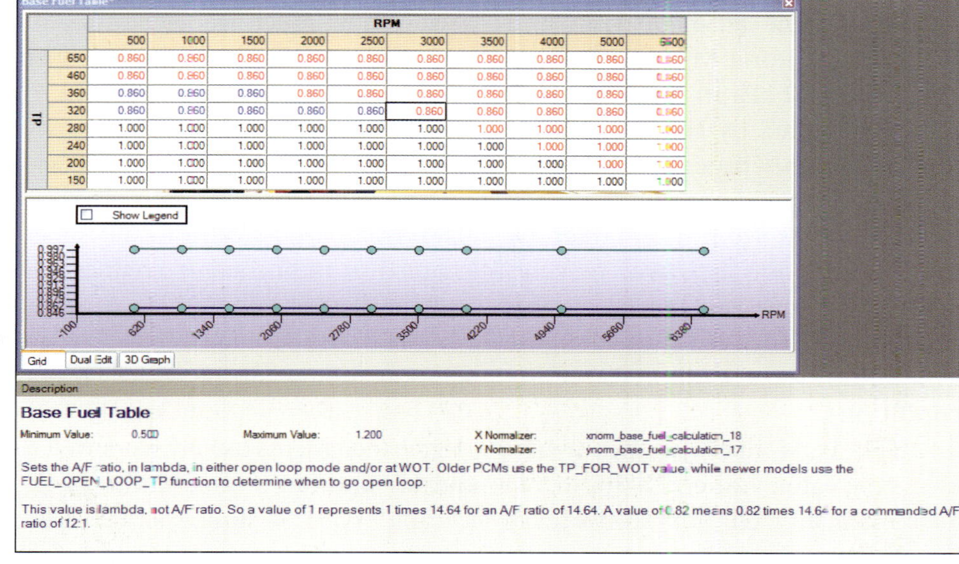

Base Fuel Table

TP	RPM									
	500	1000	1500	2000	2500	3000	3500	4000	5000	6000
650	0.860	0.860	0.860	0.860	0.860	0.860	0.860	0.860	0.860	0.860
460	0.860	0.860	0.860	0.860	0.860	0.860	0.860	0.860	0.860	0.860
360	0.860	0.860	0.860	0.860	0.860	0.860	0.860	0.860	0.860	0.860
320	0.860	0.860	0.860	0.860	0.860	0.860	0.860	0.860	0.860	0.860
280	1.000	1.000	1.000	1.000	1.000	1.000	1.000	1.000	1.000	1.000
240	1.000	1.000	1.000	1.000	1.000	1.000	1.000	1.000	1.000	1.000
200	1.000	1.000	1.000	1.000	1.000	1.000	1.000	1.000	1.000	1.000
150	1.000	1.000	1.000	1.000	1.000	1.000	1.000	1.000	1.000	1.000

Show Legend

Grid | Dual Edit | 3D Graph

Description

Base Fuel Table

Minimum Value:	0.500	Maximum Value:	1.200	X Normalizer:	xnorm_base_fuel_calculation_18
				Y Normalizer:	ynorm_base_fuel_calculation_17

Sets the A/F ratio, in lambda, in either open loop mode and/or at WOT. Older PCMs use the TP_FOR_WOT value, while newer models use the FUEL_OPEN_LOOP_TP function to determine when to go open loop.

This value is lambda, not A/F ratio. So a value of 1 represents 1 times 14.64 for an A/F ratio of 14.64. A value of 0.82 means 0.82 times 14.64 for a commanded A/F ratio of 12:1.

The target air/fuel ratio is set to a constant value in all areas where the engine may operate during WOT. This helps reduce the confusion associated with trying to hit a moving target when correcting the airflow model.

A/D Counts	Previous MAF Value	Measured Lambda	RPM	Target Lambda	Correction Factor	New MAF Value
111.83	0.34					
139.23	0.51					
162.13	0.68					
198.77	1.03					
228.98	1.35					
254.55	1.72					
276.72	2.10					
295.05	2.46					
332.19	3.25					
364.34	3.95					
393.27	4.78					
418.69	5.57					
441.58	6.37					
460.23	7.11					
478.55	7.95	0.78	1500	0.78	1.00	7.95
496.02	8.75	0.78	1720	0.78	1.00	8.75
526.16	10.36	0.78	1850	0.78	1.00	10.36
546.06	11.52	0.77	2250	0.78	0.99	11.37
574.56	13.37	0.74	2440	0.78	0.97	13.02
629.69	17.21	0.75	2800	0.78	0.96	16.55
678.72	21.11	0.77	3100	0.78	0.99	20.84
700.86	23.12	0.77	3340	0.78	0.99	22.82
760.72	29.02	0.78	3670	0.78	1.00	29.02
829.19	36.70	0.80	4420	0.78	1.03	37.64
859.13	40.49	0.80	4650	0.78	1.03	41.52
912.47	48.03	0.78	5500	0.78	1.00	48.03
981.30	58.96	0.74	6100	0.78	0.97	57.45
1023.98	67.17	0.74	6800	0.78	0.95	63.73

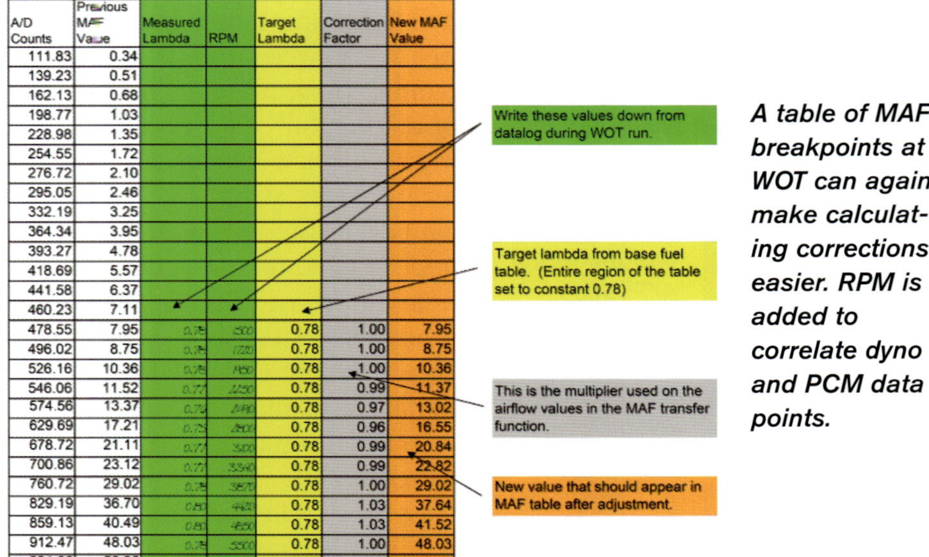

A table of MAF breakpoints at WOT can again make calculating corrections easier. RPM is added to correlate dyno and PCM data points.

the MAF transfer function, we might see a value of 30.0 lbs/min airflow at 3.2v. The airflow correction is applied to find the new airflow value to be entered into the MAF transfer function in the PCM.

$$\text{New MAF value} =$$
$$(30.0 \text{ lbs/min}) \times 1.058 \approx 31.8 \text{ lbs/min}$$

Speed density applications simply point to a cell in the volumetric efficiency table at 4,000 rpm and the corresponding MAP value (usually 95 to 100 kPa in this case) that can be adjusted by the same 1.058 multiplier. This process is repeated at all points where the delivered lambda varies from the target by more than one or two percent. Within a couple pulls at WOT, this process should bring the engine into almost exact desired fueling conditions.

The PCM likely rounds whatever new value you enter slightly to a step size allowed in its software, so do not be alarmed if it is not exact. The point is to get the airflow model as close to reality as possible. Most PCMs have small enough step sizes to get within 1% accuracy if you have the precision and patience.

Once airflow has been accurately mapped all the way up, the true search for power can begin. New target ratios can now be set in the PCM and verified on the dynamometer. The leaner the mixture, the closer to the knock limit the engine gets. For forced induction applications, the line between best power and detonation can be very thin. It is up to the calibrator to decide exactly how much leaner he is confident in letting a certain engine operate. I typically aim for $\lambda \approx 0.87$ (12.7:1 a/f) for naturally aspirated engines, $\lambda \approx 0.78$ (11.4:1 a/f) for positive displacement supercharged engines, and $\lambda \approx 0.80$ (11.7:1 a/f) for centrifugally supercharged or turbocharged engines, but individual cases vary.

With the desired lambda reached, ignition timing can be added to find the best power without knock. Two-degree increments usually work well to quickly find out if more timing helps. During WOT testing, actual spark advance should be recorded as well as knock sensor signal, if available. The OEM knock sensor can be a very useful tool in determining the maximum allowable advance for an engine, and should be used whenever possible.

Most OEM knock routines are very accurate, just slow to recover. OEM knock strategies often aggressively retard timing as soon as activity is detected from the knock sensor. For performance applications, it can be beneficial to reduce the attack rate of knock retard as well as increase the recovery rate that increments timing back to normal. Even though spark advance tries to return to the desired value sooner, a repeat offense simply starts the routine over with another round of spark retardation. By not changing the actual threshold, this technique preserves engine protection while reducing the duration of knock strategy intrusion on the driving experience.

In some rare instances, the sensitivity of the OEM knock sensor may falsely trigger spark retard. Changes to valvetrain components such as the addition of solid roller camshafts or an exhaust pipe striking the vehicle body can transmit noises in the same frequency range as knock to the sensor. Any interference issues should be mechanically fixed to allow normal operation, but valvetrain noise issues may need to be accepted as a necessary evil. If false knock sensor input persists in the absence of actual knock, the knock routine should be disabled. This can usually be done by setting a maximum allowable authority for knock retard to zero degrees.

Low amounts of spark advance at high engine speeds often lead to high exhaust gas temperatures because part of the charge is often still burning when the exhaust valve opens. Increasing ignition lead forces combustion to happen earlier inside the cylinder and reduces exhaust temperatures. If exhaust gas temperatures are high enough to require advancing the timing, the knock threshold may limit total advance. The solution is to richen the target lambda to reduce burn temperature and allow the increased timing lead.

Wide-open throttle spark conditions are especially prone to misfire. During misfire, the spark event simply never sets off the chain reaction that creates power. When engine loads are highest, the mixture density inside the combustion chamber is also greatest. These dense mixtures, though packed with tremendous energy potential, make it more difficult for the spark to jump the gap between electrodes. The first solution is to reduce the gap size, making for an easier strike with the same electrical potential. It is desirable to operate the engine with the widest spark plug gap permitted by normal operation in order to start combustion from more spark surface area. Sometimes the inability of the spark to jump this gap forces the reduction of its size. A compromise is made to burn efficiency in order to support higher power levels. Most naturally aspirated engines operate with a normal spark plug gap of 0.050 inch to 0.080 inch. Forced induction engines may need gaps as small at 0.028 inch to operate properly. If an engine misfires under heavy load with gaps any smaller than this, another solution should be found to preserve good operation at light loads.

The second solution to misfire is to increase the electrical potential (voltage) of the ignition system. This can be done by increasing the size of the ignition coil or input voltage to the coil itself. There are many aftermarket ignition systems that can be easily retrofitted to most applications. One of the primary benefits of multiple-coil ignition systems can

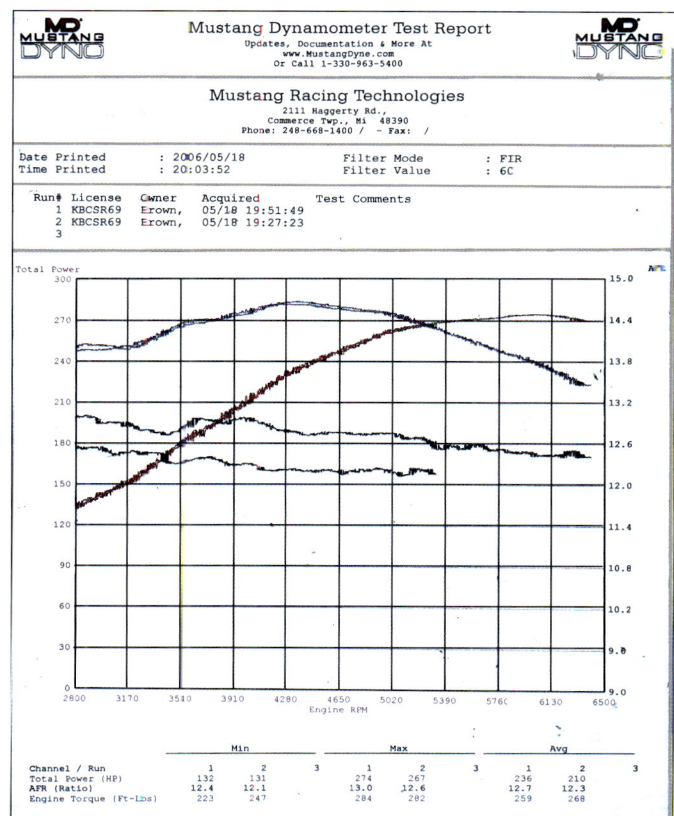

Mustang Dynamometer Test Report
Updates, Documentation & More At
www.MustangDyne.com
Or Call 1-330-963-5400

Mustang Racing Technologies
2111 Haggerty Rd.,
Commerce Twp., Mi 48390
Phone: 248-668-1400 / - Fax: /

Date Printed	: 2006/05/18	Filter Mode	: FIR
Time Printed	: 20:03:52	Filter Value	: 6C

Run#	License	Owner	Acquired	Test Comments
1	KBCSR69	Brown,	05/18 19:51:49	
2	KBCSR69	Brown,	05/18 19:27:23	
3				

Total Power — Engine RPM

Channel / Run	Min 1	Min 2	Min 3	Max 1	Max 2	Max 3	Avg 1	Avg 2	Avg 3
Total Power (HP)	132	131		274	267		236	210	
AFR (Ratio)	12.4	12.1		13.0	12.6		12.7	12.3	
Engine Torque (Ft-Lbs)	223	247		284	282		259	268	

be seen in the form of better WOT coil saturation. This is due to the increased time each coil has to recharge before striking when only supplying a portion of the engine's total ignition needs.

Additionally, insufficient spark advance can have the burn taking place too late as the piston is already rapidly moving downward, expanding cylinder volume. The resulting unstable burn yields poor power delivery and often feels exactly like a misfire to the driver. A heavy "bucking" under load that is not accompanied by obvious sounds of detonation may be improved by advancing spark lead. Remember that misfire conditions appear as lean to a wideband oxygen sensor due to the large amounts of unburned oxygen present in the exhaust.

The balance between lambda and spark lead can yield similar

Changes in the delivered air/fuel ratio can be measured with a wideband oxygen sensor on the dyno and used to determine best output. Notice the minimal change in power with change in lambda between the runs.

power levels out of the same engine with two different calibrations. It may be that $\lambda = 0.90$, 27 degrees of advance, $\lambda = 0.87$, 28 degrees of advance, and $\lambda = 0.84$, 32 degrees of advance may make the same power on an engine with obviously different fuel consumption rates. The difference is in how close each calibration rests to the knock limit and maximum capacity of the fuel system. Special care should be taken when considering possible loading and environmental changes the car may see. This is to make sure some degree of safety margin is left in the tune to prevent engine damage under unforeseen circumstances.

There are a couple of serious caveats to consider during WOT, high speed airflow modeling. First, we have made the assumption that fuel flow is known and consistent. If the fuel injectors are forced to go into static flow and the maximum limit of fuel mass delivery is reached, the engine leans out as airflow increases. It is up to the calibrator to check injector duty cycle and make sure that this is not the cause of a lean condition under load.

Insufficient fuel delivery can also result from a loss of fuel pressure at the rail. This is usually a result of insufficient fuel pump flow, but sometimes results from fuel delivery lines undersized for the necessary volume.

The second serious consideration for WOT airflow modeling is sensor range. As mentioned earlier, some MAF sensor ranges may be exceeded by engines that make large amounts of power. The same can happen when adding boost to a speed density system where the MAP sensor only has one bar of range. The ideal solution is to select a new sensor with a wide enough range to prevent "pegging" at high flow rates and input this new transfer function into the PCM. Always allowing the PCM to know exactly how much airflow the engine is developing allows it to maintain the correct lambda even if airflow changes slightly above what would have been the previous measurable limit.

For example, adding a supercharger to an engine increases power 50% above stock, and the MAF is replaced with a new unit of similar range. Running this engine at WOT on a cold day with denser air can yield another 5% increase in air mass entering the engine. If the MAF sensor is able to register this increase, all is well and the delivered air/fuel ratio never changes. If the PCM does not have any way to see this increased air mass delivery, it assumes that nothing changed and fuel delivery remains constant. This is a doubly dangerous situation, as the actual air/fuel ratio will lean out, bringing the engine closer to its detonation threshold, and more total power is being made, placing extra stress on internal components at the same time.

That same supercharged engine may also encounter belt slippage that reduces the air mass being force fed into the manifold. If the MAF and MAP sensors are both within range at this time, delivered lambda remains as commanded. If the engine is beyond the range of its sensors, this slippage results in a rich condition (often accompanied by too little spark advance for the reduced load), leading to further reduced power and possibly misfire.

If it is not possible to change the MAF or MAP sensor to accurately measure exact airflow at WOT, a few tricks can be applied to compensate. Admittedly, these are definitely not the ideal way to calibrate a car, but they work when the limits of the hardware prevent the correct solution from being an option.

Since most PCMs have either a separate table for wide open throttle fuel and spark, or at least a separate row/column in the base maps, WOT can be calibrated independently from the rest of engine operation. These are sometimes labeled as WOT maps, or "power enrichment," and may be triggered by TPS input or load calculation. Under ideal calibration conditions they are merely set to the exact desired lambda. When the range of the MAP or MAF sensor is exceeded and the engine runs leaner as airflow increases, this effect can be counteracted by commanding a richer than actual mixture in these tables or cells. The MAF is mapped and WOT lambda is set normally up until the speed that coincides with maximum MAF value. Beyond this point, commanded lambda values are set intentionally richer to maintain a constant delivered air/fuel ratio. This speed versus enrichment method works well as long as the amount of error remains consistent. Just like any other time with a sensor out of range, slight changes to the actual value that are still out of this range are not registered by the PCM. This means that any increase in airflow (colder inlet air temperatures, higher boost pressure) will lean out the actual delivered ratio. The calibrator must plan ahead and decide how much safety margin to leave in the tune for just such occasions that inevitably happen. Running slightly richer seldom costs too much power, but it can save an engine from destruction the first time the driver floors the gas on a cool autumn night.

Load calculations are also affected when the MAP or MAF sensor range has been exceeded. When the PCM continues to read constant airflow at the sensor's range with increasing engine speed, it calculates lower engine load values as speed increases. This may result in the engine shifting downward in the load maps that control commanded lambda and spark. When calibrating around a sensor in such a condition, it is important to review the calculated engine load at WOT and ensure that fuel and spark values in the appropriate cells have been adjusted to continue to yield the actual desired engine operation. It would not be surprising if the PCM for a supercharged engine with a pegged MAF only calculated 70% load at redline of 6,000 rpm at full boost. The 70% load fuel and spark targets simply need to be adjusted to accommodate. This has little chance of negatively affecting drive quality since the only time the engine spends any time at 6,000 rpm is during either hard acceleration or deceleration. Having less- than-ideal spark lead and rich fueling at "cruise" loads, and very high engine speeds most likely won't ever be noticed by most discriminating drivers.

POLISHING A SCULPTURE

At this point, most of the work is complete. If all the steady state values are correct, the vehicle should operate fairly well. The final phase of calibration is improving drivability. This should be thought of as finely polishing a sculpture. Where we were using a chainsaw earlier, we are now using sandpaper. There shouldn't be any need for drastic changes anywhere. If large changes are needed at this point, chances are that the earlier modeling is not correct or the vehicle has hardware issues.

Integrating Fuel Maps

During the mapping phase, lambda was set to 1.0 for all but wide-open throttle. During WOT testing, an ideal lambda was found that makes the best power without knock. It now becomes time to blend the map for smooth transition between cruise and power. For the majority of the base fuel map, lambda remains around 1.0 for emissions and fuel economy. If the engine is to operate at $\lambda = 0.85$ at 90% load at WOT, some sort of blending needs to happen between there and the cruising region. Since light acceleration is usu-

ally acceptable at $\lambda = 1$, the enrichment should usually start above 50 to 60% load. Try even increments between 1.0 and 0.85 for the cells leading toward stable WOT operation. If engine or exhaust temperatures get too hot when the engine is held in this transition region, more enrichment can be added. If response is still crisp and temperatures are acceptable, slight enleanment in the transition area may help improve fuel economy. This map should look like a smooth function when viewed in 3D. Sharp breaks in commanded lambda usually indicate a cover-up for some other issue such as insufficient acceleration enrichment or improper ignition tuning.

For supercharged engines operating above 100% load, a reasonable estimate of target lambda should be used for the same engine without boost for the 70 to 100% load regions. This means that a supercharged engine operating at $\lambda = 0.80$ at WOT (130% load, 150 kPa) should still have a target of $\lambda = 0.87$ at 80% load (100 kPa). It is recommended that target lambda remain fairly consistent under boost, but there should still be a smooth ramping of fuel values

between cruise and 100%. In some cases, it may be desirable to tune the base fuel map for extra enrichment above a standard boost level. This can be done by ramping in more enrichment relative to load (in the base commanded fuel map) above the calibrated WOT line to provide more cooling and knock resistance during over-boosted conditions.

Integrating Spark Maps

Much the same way the base target fuel map was blended, a similar operation must be done with the base spark map. The areas between cruise and WOT should be smoothly blended. This table should also look like a smooth function if viewed in 3D. Remember that supercharged engines usually tolerate the same spark advance as their naturally aspirated counterparts at approximately 80% load. If the engine is supercharged with a lower static compression ratio than the typical aspirated counterpart, more timing is necessary at lower loads to compensate for the reduced cylinder pressure. The idea is to run the engine close to MBT timing in these transitional regions in order

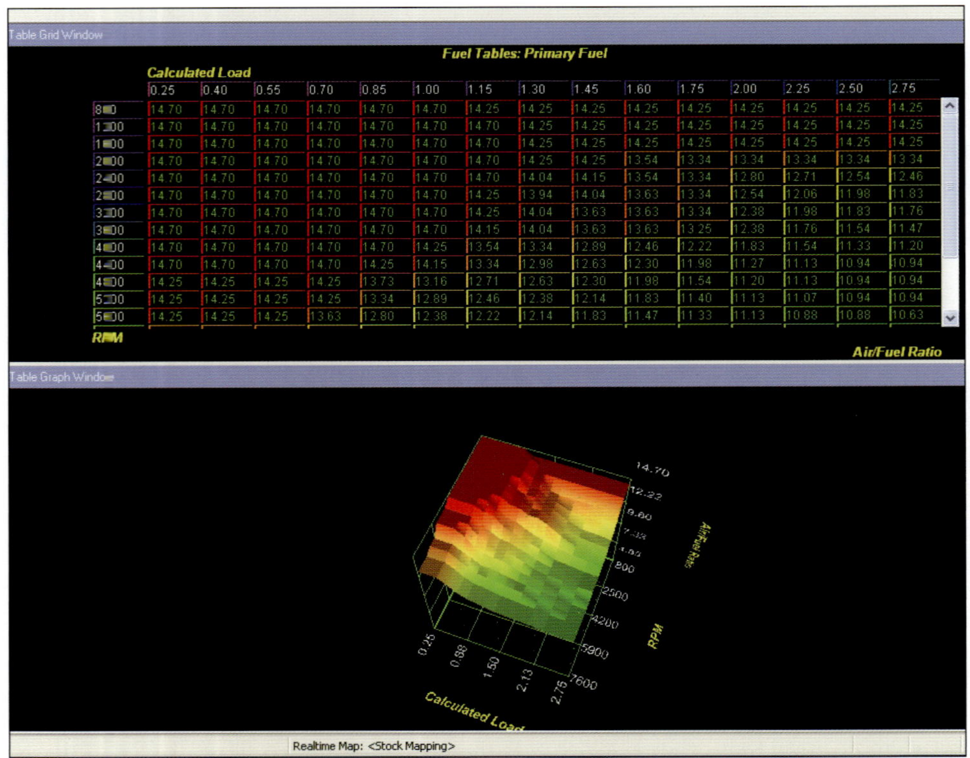

RPM	0.25	0.40	0.55	0.70	0.85	1.00	1.15	1.30	1.45	1.60	1.75	2.00	2.25	2.50	2.75
8□0	14.70	14.70	14.70	14.70	14.70	14.70	14.25	14.25	14.25	14.25	14.25	14.25	14.25	14.25	14.25
1□00	14.70	14.70	14.70	14.70	14.70	14.70	14.70	14.25	14.25	14.25	14.25	14.25	14.25	14.25	14.25
1■00	14.70	14.70	14.70	14.70	14.70	14.70	14.70	14.25	14.25	14.25	14.25	14.25	14.25	14.25	14.25
2■00	14.70	14.70	14.70	14.70	14.70	14.70	14.70	14.25	14.25	13.54	13.34	13.34	13.34	13.34	13.34
2□00	14.70	14.70	14.70	14.70	14.70	14.70	14.70	14.04	14.15	13.54	13.34	12.80	12.71	12.54	12.46
2■00	14.70	14.70	14.70	14.70	14.70	14.70	14.25	13.94	14.04	13.63	13.34	12.54	12.06	11.98	11.83
3□00	14.70	14.70	14.70	14.70	14.70	14.70	14.25	14.04	13.63	13.63	13.34	12.38	11.98	11.83	11.76
3■00	14.70	14.70	14.70	14.70	14.70	14.70	14.15	14.04	13.63	13.63	13.25	12.38	11.76	11.54	11.47
4■00	14.70	14.70	14.70	14.70	14.70	14.25	13.54	13.34	12.89	12.46	12.22	11.83	11.54	11.33	11.20
4□00	14.70	14.70	14.70	14.70	14.25	14.15	13.34	12.98	12.63	12.30	11.98	11.27	11.13	10.94	10.94
4■00	14.25	14.25	14.25	14.25	13.73	13.16	12.71	12.63	12.30	11.98	11.54	11.20	11.13	10.94	10.94
5□00	14.25	14.25	14.25	14.25	13.34	12.89	12.46	12.38	12.14	11.83	11.40	11.13	11.07	10.94	10.94
5■00	14.25	14.25	14.25	13.63	12.80	12.38	12.22	12.14	11.83	11.47	11.33	11.13	10.88	10.88	10.63

A smooth commanded air/fuel ratio table is shown using the Cobb ProTUNER software. This table shows the stock values for a Subaru STI that have a progressive enrichment as speed and load increase.

to extract maximum efficiency without knock.

Drivability improvements are done largely by adjusting the transient controls to provide smooth changes between steady state conditions. The first transient to be refined is acceleration enrichment (AE). Adjustment of this parameter can be highly subjective. The object is to supply just enough extra enrichment to allow smooth transition to power without dumping excess fuel and hurting economy and emissions. When calibrating AE, it is helpful to think about how much of the wall film is being evaporated due to the increased airflow and add just enough fuel to keep τ constant. A quick stab at the throttle pedal momentarily shows a lambda leaner than the commanded value for high load. The acceleration enrichment multiplier should be increased until the wideband goes directly to the targeted steady state lambda at the higher load. The larger the camshaft, the more aggressive this function needs to be. Acceleration enrichment is usually adjustable relative to throttle position, so this should be checked with tip-ins starting at idle, part, and medium load. More AE is necessary at closed throttle because of the large amount of area increase when rotating the blade here. Fine-tuning of this function should be done under normal driving conditions with typical tip-in rates. If the vehicle stumbles, then picks up and goes, more AE is likely needed and the wideband should show lean during the stumbling. Adding fuel makes tip-in smoother, but too much can foul plugs, hurt economy, or increase emissions.

Even more accurate calibration of the AE can be done with high speed logging of wideband lambda and target ratio. While performing a tip-in with a known load change, but at a constant target lambda, AE requirements can be shown. This is best performed at part load where target $\lambda = 1$. A change from 15% load up to 50% load should not incur a change in the delivered lambda. By logging the actual lambda during this transition, AE requirements can be determined. If the wideband shows a momentary lean condition during load change, more acceleration enrichment is needed. Likewise, momentary rich conditions indicate excessive AE. This process assumes that all static airflow and fuel mapping has already be done with a high degree of accuracy. In OEM level calibrations, this process is repeated for a large array of speed, load, and temperature points to provide the best possible lambda control for emissions under all possible conditions.

Tip-In Ignition

While acceleration enrichment helps to ensure proper fueling during tip-in, it may sometimes be necessary to take additional measures to prevent knock or driveline noise. Burst knock is a phenomenon that can occur during sudden increases in engine load. The rapidly rising cylinder pressures may lead to knock under what would otherwise be stable combustion at steady state operation. Even without knock, the sudden onset of engine torque may lead to driveline noise as the lash is quickly taken up. Many OEMs intentionally reduce the available torque onset to keep noise low or prevent driveline component failure. Both of these functions are typically controlled

This ECU is from a Cadillac Northstar engine. With an advanced circuit board design, the majority of its surface area is filled by the actual wire harness connections.

by momentarily reducing spark advance. If instant throttle response is desired, the spark retardation can be set to zero. Added acceleration enrichment can be used to quench minor knock while retaining full timing and better torque. If no specific function exists in the PCM being used and tip-in retard is desired, the appropriate cells in the base spark map can be reduced. Since the engine is not likely to spend any time cruising above ~50% load at low speed, it is usually safe to reduce timing below MBT here to soften tip-in or reduce burst knock.

Deceleration

Dashpot adjustment can be checked at this point by simply lifting off the gas pedal from an elevated engine speed. The same should also be done by cruising at a steady speed and pushing in the clutch or shifting to neutral. The object is to find the balance between hanging at the elevated speed for too long after lifting versus dropping too quickly past idle speed and stalling. If the engine speed

hangs, decrease the IAC position at the same engine speed or increase the decrement rate of the IAC position. If deceleration tends to drop right past idle speed (and the closed throttle position at idle is correct), the commanded IAC position at higher engine speeds should be increased or the decrement rate should be reduced. Initial adjustments to this function should be done in 10 to 20% increments to see enough difference in actual performance.

Some difficult engine combinations may not allow for a quick descent immediately to idle speed. In these cases, the IAC position versus engine speed knee point can be moved to a few hundred RPM above idle. This allows for engine speed to drop to a lower intermediate speed where stalling is less likely and the idle controls can softly move toward the desired target with a softer landing.

Deceleration enleanment (DE) can be adjusted next. Some PCMs have tables specifically for deceleration enleanment. These tables allow for the adjustment of the turn on/off points based on engine speed. These should be set well above idle to avoid stalling. The larger the camshaft overlap and duration, the bigger the gap between idle and DE threshold should be. Many stock engines that idle around 650 rpm can tolerate DE all the way down to 1,000 rpm or so. A modified engine with high compression and overlap may drive better by only allowing DE above 3,000 rpm.

On PCMs without such tables, the same effect can be accomplished by adjusting the row or column of base fuel map cells for high engine speed and low load (high vacuum). Since the only way the engine could ever operate in these cells would be to have the throttle closed at high speeds, this fills

the definition of deceleration. Setting target lambda to an extremely lean value commands lean operation in these conditions. A value of 0% VE also forces fuel shutoff and can be used in simpler speed density systems to achieve the same result. Since normal combustion is not present and any actual loads are very small, spark advance values for these cells can also be relatively high.

DE feel to the driver is tied to dashpot as well. Dashpot should be the primary control of engine torque and speed during deceleration with DE only activating in areas where very fast drops in engine speed are desired. If a gentle drop in engine speed is desired at any point, DE should not be used since it compromises the engine's ability to produce consistent torque output.

Closing the Loop (or not?)

After airflow has been mapped correctly under most conditions, closed loop operation can be considered. All modern OEM systems use closed loop operation to compensate fuel delivery for weather conditions, part wear, and changes in tolerances. This operation is a very reliable method of keeping the engine in stoichiometric operation as long as the system inputs (primarily from the HEGOs) are reliable.

The timing of the HEGO signal is carefully adjusted in the factory PCM. The PCM makes adjustments to delivered fuel based upon the amount of "flight time" or transport delay between the exhaust valve opening and HEGO sensor measurement. When an engine is operating very close to stoichiometry and only small adjustments are being made, it is important to adjust in the correct

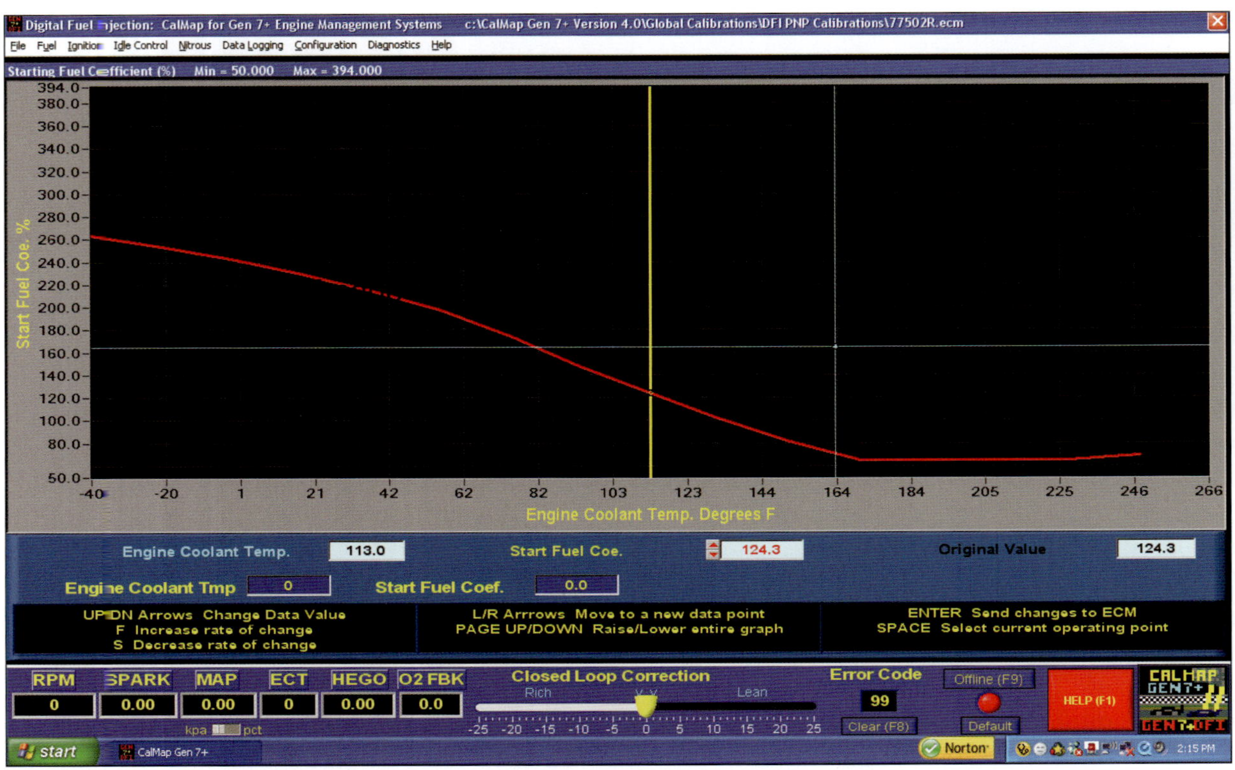

Only after the base calibration is done can we progress to cold start tables. At this point, you should resist the urge to adjust the base fuel tables to correct for any roughness.

direction at the proper time to avoid unstable correction conditions. If modifications to the vehicle have forced moving the HEGOs further away from the cylinder head, it may be necessary to adjust the transport delay functions in the PCM to compensate. Otherwise the fast adjustments being made by the PCM during closed loop operation may become out of phase with the HEGO feedback, forcing the learned values to diverge from ideal. Adjusting the transport delays can put these back in synchronization, improving fuel economy and emissions. Typically, a change from a short manifold to long-tube headers adds about 20 to 30% to the transport delay times due to sensor location.

Some mild performance upgrades may prevent proper closed loop operation at cold temperatures. OEM systems have temperature-based triggers that can be adjusted upward to reduce unwanted adjustments during warmup.

In cases where the sensor is moved too far away from the head to remain hot even with its internal heater, the reduced accuracy may prohibit reliable closed loop operation at low speed. Additionally, large camshafts with significant overlap may experience too much charge wash at low speed for the HEGO to determine an accurate exhaust only signal. In some cases, this can be remedied by setting a closed loop operation threshold to an engine speed high enough to maintain good feedback from the HEGOs. If this is not possible, full-time open loop operation may be the only other choice.

If the airflow mapping has been done accurately, there should be very little closed loop correction needed anyway. This is the part of the process that reveals just how precise one was during the airflow mapping. Ideally, the airflow modeling was performed correctly for all speed and load conditions, and closed loop corrections are less than about 3%. Nothing should really change too much at this point. If closed loop operation suddenly causes a shift of 20% in fueling, there may still be more work to be done on the base tables. This is the time to stop and take a closer look at your previous work.

Choosing Cam/Runner Timing

For engines with variable valve timing or intake manifold geometry, measurements should be taken under most possible conditions. This usually means making a power sweep with the cams fixed to minimum and maximum advance or intake port flaps open and closed. By overlaying the resulting torque curves an intersection point can be found. This is usually the best point for switchover between the two conditions.

For infinitely variable systems, this process can become more complex. A good starting point is to perform sweeps for minimum, maximum, and mean positions and compare these to the OEM calibration if available. The OEMs typically hold variable cam timing adjustment to a later point to preserve emissions during low speed test cycles. More torque can often be found by comparing the OEM response curve to a locked sweep to find a better transition.

If the camshaft itself has changed in an infinitely variable system, be prepared for a longer dyno session to find the ideal crossover points. Start with an OEM curve and work from there.

Choosing a Shift Point

By examining the WOT power curves, an ideal transmission shift point can be determined. Many OEM automatic-equipped PCMs also control shift patterns. While the focus of this book is not transmission calibration, engine output plays a significant part in WOT shift patterns. The method of picking this shift point is similar for either automatic or manual transmissions.

If calibrated correctly, most engines exhibit a parabolic curve for both power and torque at WOT. There will be a high-speed point on the curve at which the engine makes peak power. Beyond this point, power drops as speed increases. How fast this drop occurs determines where the shift point should be.

If an engine with a redline of 6,600 rpm makes 300 hp at 6,000 rpm and drops to 260 hp by 6,300 rpm, there is no reason to shift much beyond 6,100 rpm. Shifting before actual redline allows the engine to enter the next higher gear near the meat of the torque curve. However, if another engine makes the same 300 hp at 6,000 rpm but continues to make 295 hp at 6,500 rpm, the shift point should be much later. If this second engine makes significantly less torque than the first, this is even more reason to delay the shift point as much as possible, perhaps close to redline.

Second to detonation, RPM is the largest strain on an engine. If peak power occurs at a lower engine speed, waiting to shift until redline only serves to reduce engine life and slow the vehicle down while making more noise. If engine power is no longer increasing, high RPM is usually not going to help.

Almost Done

Once all of this work has been completed, drive the car. This may sound obvious, but if most of the work was performed on a chassis dynamometer, the road may be different. Even a load bearing dyno is slightly different from the real world due to the added airflow across the car, radiator, and engine compartment. The unique driving style of one person may expose conditions unseen during testing. Try to drive both normally and as awkwardly as possible during the test drive to see as many real-world conditions as you can. Hold the vehicle in high gear and lug the motor with moderate acceleration to check for bucking, misfire, tip-in response, and knock. Perform one WOT pass (if it won't get you arrested) to check for knock, bucking, or unusual performance. Coast to a stop in neutral to verify dashpot calibration. Stop the car for a minute and check restart with the engine hot. If possible, allow the car to cool completely and check cold starting.

Congratulations. You have successfully calibrated like a professional.

Cars (like this one) with a larger aftermarket camshaft installed require changes to the transient fuel tables to improve drivability. A slight increase to the wall film tables will go a long way toward curing the typical tip-in stumble associated with larger cams.

FORCED INDUCTION

So far, the focus of this book has been naturally aspirated engine calibration. Although forced induction has been occasionally mentioned, it deserves some more focused attention. The aftermarket performance industry is flush with ways to add power to an engine. Few of these methods come close to the potential that forced induction offers for power increases. Many OEM vehicles are built with superchargers or turbochargers from the factory, which often leaves the door open for the enthusiast to simply increase the pressure ratio in the search for power. At the end of the day, the concerns are the same. Any time more compression happens to a gas, its temperature rises by some amount. How much temperature increase comes along with the compression is a direct result of the amount of compression happening and the efficiency with which it is being done. Not all compression methods are created equal.

It is important not to confuse manifold pressure with cylinder pressure. The two are linked, but the relationship between them is unique for each engine combination. Cylinder pressure is the primary deciding factor when considering how much less timing is needed under boost compared to the same engine under atmospheric conditions. Looking at load instead of boost gives a better representation of what the engine wants. Increases in airflow cause load shifts, not boost. Boost is just a measure of restriction between the engine and compressor. If exhaust backpressure or camshaft design is changed to increase flow, boost may actually decrease (without a change in drive/pulley ratio) while airflow and load increase. This condition may require a reduction in timing to prevent knock at the higher load even though boost has been reduced. The only way to know for sure is to test under load.

As mentioned before, extra fueling is used to improve combustion stability and reduce temperatures under high load. The increase in

The Ford Lightning is an example of a vehicle that came from the factory with some form of supercharging. The ECU has logic that can accurately calculate engine loads in excess of 100%. (Nate Tovey)

available airflow may also require revision to acceleration enrichment requirements to accommodate the higher instantaneous increases in airflow on tip-in. Supercharged engines should still be able to operate normally at low and mid load. The addition of forced induction alone does not require that the engine always be operated rich. The normal procedure for part throttle mapping remains and $\lambda = 1$ fueling for most operating conditions should still be targeted to preserve fuel economy and emissions.

Part load spark calibration is again almost identical to that of a naturally aspirated engine. MBT is not affected by the supercharger at part throttle. MBT remains the ignition target at part load. There will just have to be a blending into the new higher load regions of the map.

Centrifugal Superchargers

Probably the most widely used compressor in the performance aftermarket is the centrifugal pump. Companies such as Vortech, Paxton, and ATI produce countless systems for almost every performance vehicle on the market. These compressors are driven by the accessory drive belt, linking their impeller speed directly to engine speed. Because these compressors offer little increase in flow at low speeds, they are overdriven through both pulley ratio and internal gear sets. Still, the centrifugal pump design requires relatively high speed to reach its potential. Actual engine boost output from a centrifugal supercharger increases with engine speed. Drive losses for a centrifugal supercharger also vary depending on the load. At idle they only require a few horsepower to turn, but losses

This engine is equipped with a Paxton centrifugal supercharger. A longer accessory drive belt is used to accommodate the extra pulleys and positioning. (Nate Tovey)

increase with speed. A typical aftermarket street supercharger can take anywhere from 30 to 50 hp to drive at full load. This is very real power that must be transferred from the crankshaft to the supercharger, and it must be taken into account when calculating fuel system requirements. Luckily, the drastic increase in airflow more than outweighs the parasitic losses at high load.

As impeller speed increases, so does pumping volume and efficiency… to a point. The result is usually a wide-open throttle boost curve that increases almost linearly with engine speed. At low RPM, there is almost no added manifold pressure. Therefore, engine calibration at these speeds can be done with a leaner lambda and more aggressive spark advance below peak camshaft efficiency, just like on a naturally aspirated engine.

Even with relatively little increase in engine load at low throttle position, air is still being pumped by the compressor at low speed. This

pumping volume is still greater than the engine's airflow requirements at idle and low load, so pressure builds between the throttle blade and compressor. The added pressure on the compressor blades creates drag on the accessory drive that can make idle control difficult. A bypass valve is used to divert this pressure back to the inlet side of the compressor. The valve is actuated by a vacuum

The drive belt for the supercharger can be clearly seen here. A dedicated tensioner is also used to reduce belt slip.

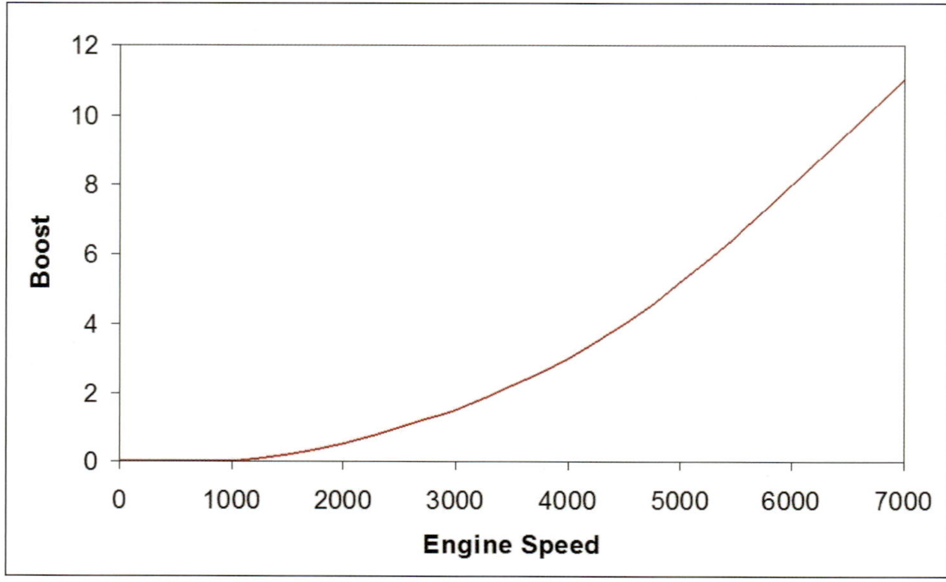

Engines equipped with centrifugal superchargers make peak boost at redline. The exponential response generates most additional loading after the engine passes peak torque. (Nate Tovey)

ciency of the centrifugal pump becomes greater.

Because of the variable load increase, a centrifugal supercharger system rated to deliver 10 psi (0.7 bar) of boost only does so at maximum speed. If this maximum speed is 6,500 rpm, there may only be 5 psi (0.35 bar) of boost present at the torque peak near 4,500 rpm. This tells us two things. First, engine load increase near peak torque (the point at which cylinder pressure would normally be highest) is significantly less than what is seen at redline. This means that less timing retard is necessary at this point in order to maintain good engine durability and knock resistance versus redline. Secondly, increasing loading means that the high-RPM pumping efficiency of the newly supercharged engine system is extended. An engine that would otherwise lose volumetric efficiency near 6,000 rpm may now continue to develop useful power to 6,500 rpm or higher. This is due to the ever-increasing manifold pressure overcoming the

diaphragm connected to the intake manifold. When the throttle is opened and manifold vacuum disappears, the bypass valve closes and forces all compressor output to be routed directly into the engine.

One word of caution for mass airflow based systems is to ensure that all flow only passes the metering element once. It is a common mistake to route bypass air back into the inlet upstream of the MAF sensor, causing double metering during bypass. This makes proper MAF calibration all but impossible. Likewise, if the bypass is placed after the MAF element, it must not vent to atmosphere. Dumping metered air back to the atmosphere results in a rich condition as the PCM assumes all metered air is actually going into the engine.

As engine speeds increase, boost pressure at WOT also increases, resulting in loads exceeding 100%. Load continues to increase all the way until redline unless otherwise restricted. Unless the drive belt slips, centrifugal

superchargers develop maximum boost at redline. Cylinder pressures should steadily increase with speed as well, with moderate increases near peak torque and substantial improvements at high speed where the effi-

This supercharged engine makes over 1,700 hp and an appropriately large amount of boost. To vent this tremendous pressure when the throttle is closed, two bypass valves are employed to reduce the load on the compressor wheel. (Nate Tovey)

loss of ram tuning effectiveness. When peak load is reached at high engine speed, more timing must be pulled to avoid knock. The increased inlet temperature that comes with boost should be compensated for in the calibration. This is done by adjusting the IAT timing function accordingly to pull timing at high temperatures. For this reason, the IAT sensor should be placed such that it always measures actual manifold air temperatures in supercharged applications.

Most efficient centrifugally supercharged engines can be safely calibrated to operate with a lambda close to that of a naturally aspirated engine for short durations. However, in the interest of durability, a slight enrichment with the resulting reduction in combustion temperatures is desirable. The hotter the actual inlet air temperatures are anticipated to be, the more enrichment should be used. I have typically targeted λ = 0.8 (11.7 a/f) as a starting point for all conditions in excess of 100% load with centrifugally supercharged engines. Some intercooled applications with lower inlet temperatures may be just fine at λ = 0.85 (12.5

a/f), but I still prefer to stay on the safe side for street-driven vehicles. On one centrifugally supercharged, non-intercooled drag race application, I found best power at λ = 0.89 (13.1 a/f)! This engine actually worked very well with no signs of knock and terrific performance, even if it was only for 10 seconds at a time. Obviously, the short duration of load in the drag racing environment allowed for the hotter burn without engine damage.

Generally speaking, charge temperatures dictate how rich the engine must be run under load to preserve durability. Since centrifugal superchargers only produce peak boost (and temperatures) for a brief period of time near redline, their need for charge cooling is not as great as other superchargers. If a charge cooler is used, target lambda can be higher. Since there are usually diminishing returns on power increase from enleanment, a slightly rich mixture can be kept with the power being found in ignition timing. The engine's knock limit is usually the limit on how much power can be made at a certain boost level.

The classic hot rodder's "blower," the Roots supercharger is a positive-displacement compressor that mounts directly to the intake manifold. (Nate Tovey)

Positive Displacement Superchargers

The positive displacement supercharger traces its heritage back to the earliest days of automotive performance. The compact design is often integrated with the intake manifold to save space, making for an attractive OEM option. Ford, GM, Mercedes, and others have employed positive displacement superchargers to increase performance in dozens of car models. Eaton, Autorotor, and Whipple manufacture these units, which are used in various OEM and aftermarket installations.

This Eaton Model 112 compressor from a 2003 Cobra is another variant of the positive displacement compressor. The throttle body and IAC motor are installed upstream of the compressor inlet. A bypass valve (arrow) is also integrated with the back side of the unit. (Nate Tovey)

The bottom side of the Eaton supercharger shows the compression rotors (arrow) behind the triangular discharge at right. The round bypass valve (arrow) can also be seen on the left.

The principle is simple. For every rotation of the positive displacement (PD) supercharger, a fixed volume of air is moved through it. When the PD supercharger is sped up with a pulley drive similar to the centrifugal supercharger, more air is moved than the engine normally displaces. The result is almost instant boost when full throttle is applied, regardless of engine speed. Because the ratio between blower displacement and engine displacement does not change, boost can remain almost constant across all engine speeds.

There are a couple different varieties of positive displacement superchargers. The most common are Roots style such as the Eaton Model 90 found on the Pontiac Grand Prix or Model 112 on the Mustang Cobra. Top Fuel dragsters also employ a much larger variant of the Roots compressor design. A more efficient variant is the Lysholm Screw compressor such as the Autorotor and Whipple designs. The increase in efficiency comes from a more axial flow through the compressor with a different blade design, yielding less heat at the same pressure ratio as well as increased flow volume.

Parasitic losses to drive a positive displacement (PD) supercharger are higher than the equivalent centrifugal supercharger. Drive losses increase with speed for PD superchargers as their efficiency drops. The caveat is that PD superchargers become less efficient as speeds increase. Attempting to overdrive a PD supercharger can often lead to immense increases in outlet temperature without the desired mass flow increase due to efficiency losses. An overdriven Eaton supercharger on a Ford Lightning can take over 80 hp to operate!

Due to the increased low-speed pumping, idle loads are more pronounced with positive displacement superchargers. An appropriate change must be made to idle speed targets and spark values to maintain stable idle when adding a positive displacement supercharger. A bypass valve is also mandatory to prevent overheating the intake charge and excessive crank loading even at idle. The throttle blade is usually positioned upstream of the compressor section, so bypass happens entirely behind both the IAC and MAF sensor. This makes the calibration of the MAF and IAC functions similar to that of a naturally aspirated car of larger displacement.

WOT loads for positive displacement supercharged engines can easily exceed 100% even at speeds just above idle. This makes low speed spark mapping more critical than with a centrifugal supercharger. The added load means that cylinder pressures rise naturally without the need for extra spark advance below peak camshaft efficiency. Off idle, high load ignition timing requirements for PD supercharged engines are usually surprisingly low.

Nearing peak torque, cylinder pressure rises rapidly, which requires significantly less ignition lead. Unlike centrifugal superchargers, PD units have full boost at the same point as peak camshaft efficiency, leading to even higher cylinder pressures at the same peak boost levels. This added cylinder pressure translates to a lower ignition advance at the knock limit near peak torque.

After peak torque, when camshaft efficiency decreases, cylinder pressures begin to drop again in a similar manner to naturally aspirated engines. Obviously there is still boost, but timing can still be added to make up for the decreasing engine efficiency at high speed. Although timing does not approach the same total values as a naturally aspirated engine at WOT, the increase with speed after peak torque is similar.

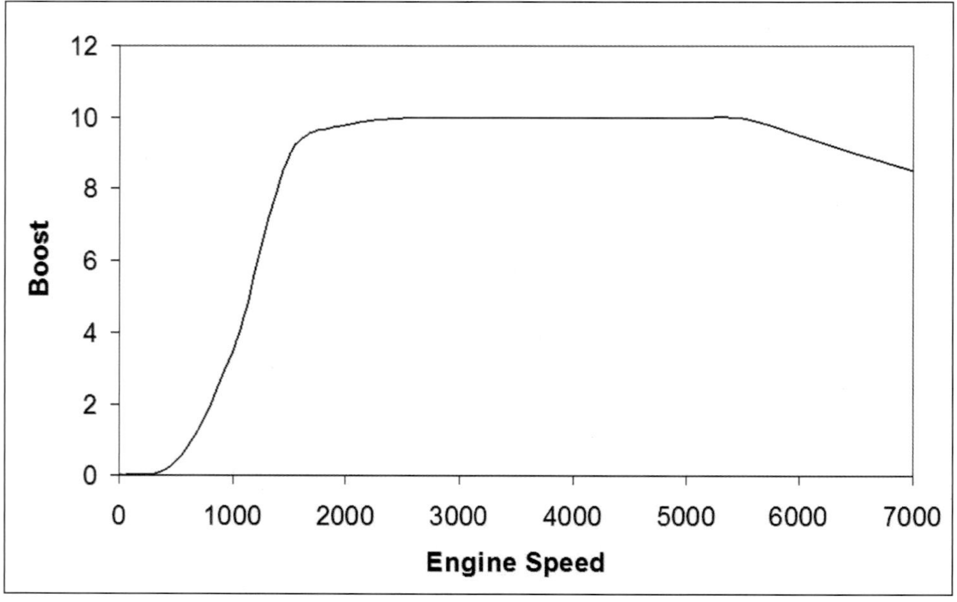

Engines equipped with positive displacement superchargers make peak boost very early. The relatively flat response generates most additional loading long before the engine reaches peak torque. (Nate Tovey)

The core of the water-to-air cooler acts much like a radiator in reverse. Coolant is passed through the ports (arrow) at right while the compressed charge flows across the fins from the top. (Nate Tovey)

Most OEM installations of positive displacement superchargers include a water-to-air charge cooler. Tight packaging requirements often dictate that these be worked into the intake manifold as an integral part. The intake runners (arrows) can be seen along the side of the plenum. (Nate Tovey)

impact on power output, but really helps stave off knock. The richer mixture burns cooler in the chamber, but ignition timing can also be increased slightly to maintain stable combustion and good power. This usually gives the best overall safety margin without hurting power. Some properly intercooled positive displacement engines are perfectly fine at $\lambda = 0.85$ (12.5 a/f), but little power is usually lost with some added enrichment that can go a long way toward avoiding knock with the occasional heat soak or tank of bad fuel.

Turbochargers

Turbochargers are the other compressor so often embraced by OEMs. A turbocharger is not driven directly by the crankshaft, but rather by the otherwise unused heat and velocity of the spent exhaust gases. This indirect coupling gives the turbocharger some unique performance characteristics.

The first difference is parasitic loading. Since the turbine section harnesses the power of the exhaust gases, it presents almost zero load to the engine at idle and part throttle. This makes idle and part-throttle calibration identical to that of naturally aspirated engines.

The boost curve is the second, more noticeable difference. It is the energy in the exhaust stream that powers the turbocharger. Placing the turbine section in this flow of hot gases turns heat and velocity of the gas flow into mechanical work that drives the compressor. The more exhaust energy that is available, the more compressor energy that is available. The buildup of this energy to the point where the compressor flows more air than the engine would on its own is known as "spooling." Many

Practical engine speed ranges are usually not increased as a result of adding a positive displacement supercharger.

Because of the tendency to run much warmer than their centrifugal counterparts, PD superchargers require additional help cooling the intake charge. Just like with centrifugal superchargers, inlet temperature should be monitored after the compressor to allow for proper adjustments to ignition lead or fueling. Most OEMs use the preferred method of charge cooling with an intercooler to reduce inlet temperatures. A water-to-air cooler can often be packaged as part of the supercharger and intake manifold assembly, saving precious underhood space.

If intercooling is not possible, extra fuel can be used to accomplish the necessary combustion temperature reductions. To keep combustion temperatures in check, I usually run a positive displacement supercharger richer than its centrifugal counterpart. A starting target of $\lambda = 0.78$ (11.4 a/f) usually works best. Again, the slight increase in fuel usually has little

Exhaust driven turbochargers have long been used by OEMs. Mounted closely to the cylinder head, they make optimal use of the heat energy of the exhaust gases.

factors go into deciding at what engine speed this point occurs at WOT, primarily turbocharger sizing relative to the engine. While we do not go into detail here about how to ensure best system efficiency and quick spooling, it is important for the calibrator to keep thinking about this during the tuning process.

Spooling is dependent upon having enough exhaust gas energy (either heat or velocity) to drive the turbine. If the turbocharger does not respond quickly enough, more exhaust energy is needed. This can be done by adding mass flow or adding temperature. Mass flow is dictated by engine hardware at WOT, so little can be done in the calibration to improve this. Temperature can be increased by retarding ignition lead or richening the mixture slightly to allow more combustion to happen outside of the chamber on the way to the turbine. If gases are still burning and expanding as they pass the turbine, more power is harnessed there rather than by the pistons. Although it results in a slight decrease in cylinder pressure, a quicker spooling and the resulting manifold pressure rise quickly offsets this. Likewise,

if spooling occurs too quickly, response is softened by advancing ignition lead or throttling airflow.

Once the turbine has harnessed sufficient power to drive the compressor to the desired manifold pressure, gases are partially routed around the turbine section through the wastegate to maintain a constant energy transfer to the turbine. The result is a boost curve sloping up to the desired level during the spooling period and holding steady until redline, assuming a proper flow match to the engine.

The turbocharger uses the more efficient centrifugal compressor design, but without the limitation of being linked to engine speed. By allowing the compressor to be driven into its peak efficiency range much sooner, overall system efficiency is increased. The power to drive the compressor registers only as an increase in exhaust backpressure to the engine. Depending on catalyst and muffler design, this may not even be any greater than the naturally aspirated equivalent. As a result, turbocharger installations often yield parasitic losses of less than 15 hp, with almost zero loss at cruise and idle. Another side benefit is that the rotating blades of the turbine section tend to break up individual exhaust pulses, quieting the exhaust note of the vehicle. A lower-restriction muffler can then be used without the usual noise penalty, again improving system efficiency.

Much like the positive displacement supercharger, WOT loads from the turbocharger represent a magnification of the engine's naturally aspirated tendencies. Although low-speed high-load operation is somewhat limited by spooling characteristics, peak boost is still often available well before peak torque. To compensate

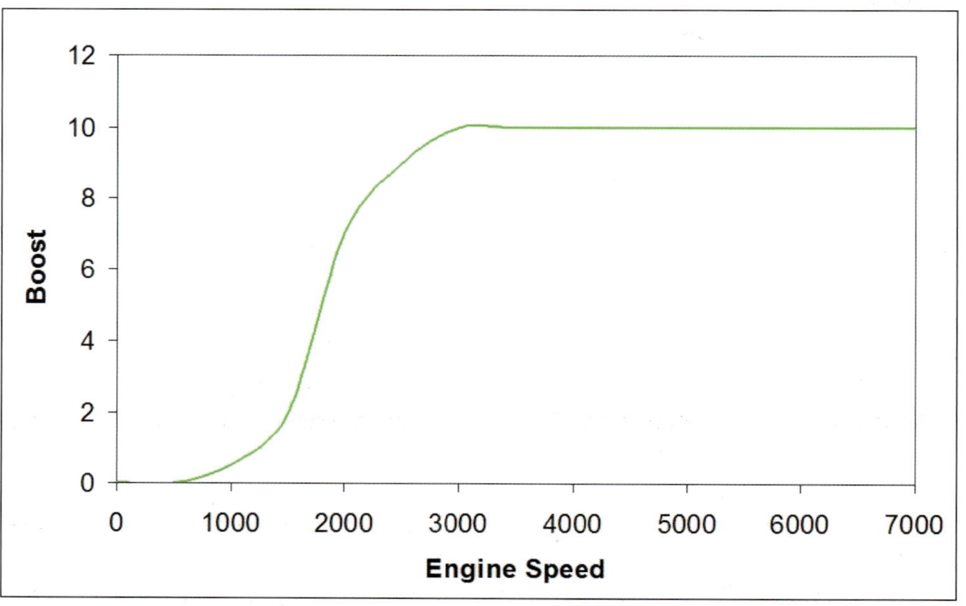

Much like the positive displacement supercharger, the turbocharger allows for full boost early in the RPM range. A small amount of "lag" is sometimes present at low speeds, but power continues very strong to redline like with centrifugal superchargers. (Nate Tovey)

The turbocharger's compressor is very similar to that of the centrifugal super-charger. The primary difference is the installation location on the exhaust manifold where the shaft is driven by hot exhaust gases instead of a belt drive.

At the drag strip, many competitors using turbo or superchargers and air-water intercoolers will fill their coolant resevoirs with ice. This helps bring the temperature of the compressed and cooled air entering the engine well below ambient for even greater power and knock resistance. (Nate Tovey)

Most OEM and aftermarket turbocharger installations include the use of an air-air intercooler. For best performance, these are often mounted in front of the radiator to maximize cooling efficiency. It is becoming more common to see these used with centrifugal supercharger installations as well. (Nate Tovey)

here, timing must once again be reduced near what would normally be peak cylinder pressure near peak cam efficiency. After this peak, timing can once again be increased with engine speed as long as inlet temperatures and cylinder pressures permit.

Unlike the positive displacement superchargers, turbochargers continue operating closer to peak efficiency all the way to redline when sized properly. This means that inlet temperatures do not increase as dramatically as with the PD superchargers at high engine speeds. Even with the increased efficiency, turbocharger systems typically include some form of charge cooling to account for the longer time duration that the engine can be exposed to boost. The effectiveness of this charge cooling largely dictates the spark advance requirements to avoid knock at WOT. Again, actual inlet temperature should be used to adjust ignition advance.

Even with significantly lower heat soak than positive displacement superchargers, turbochargers require some amount of fuel enrichment for combustion temperature control depending on intended loading. The

To control shaft speed of the turbocharger, a wastegate is used to route excess gases around the turbine. This example is built into the turbine housing for a compact installation.

This external wastegate performs much like its internal counterpart. The larger size allow for the greater bypass flow capacity and heat resistance needed on high-performance engines.

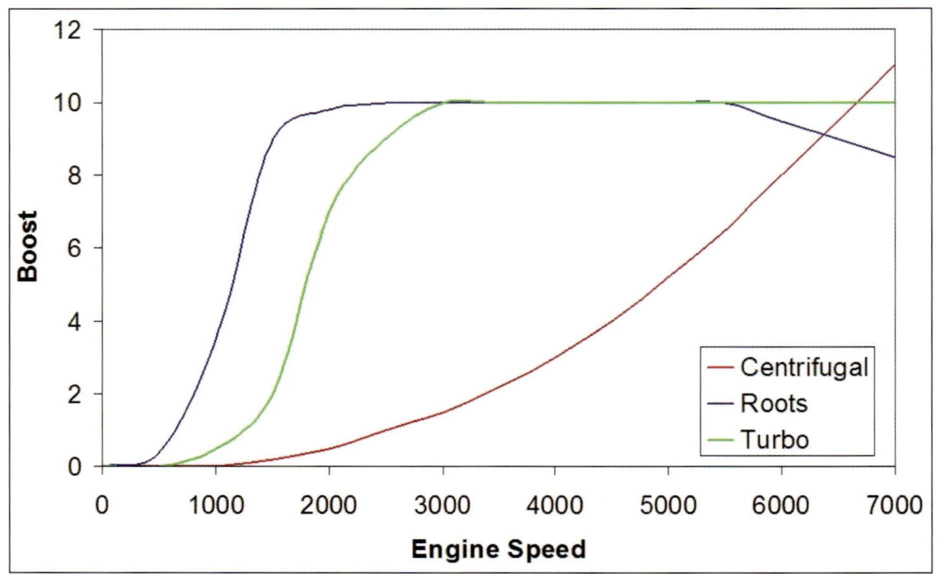

A comparison of the three primary types of superchargers is shown here. Notice the differences in boost that translate directly to possible engine load ranges at WOT. (Nate Tovey)

longer the engine is expected to perform under boost, the richer target lambda should be for durability. Drag race turbocharged engines may be set to run as lean as $\lambda = 0.85$ (12.5 a/f), but road racing or street applications are better suited to operate around $\lambda = 0.8$ (11.7 a/f) to cool exhaust valves and manifolds.

Timing should not be excessively retarded with a turbocharger at WOT. The resulting increased exhaust gas temperatures can be hot enough to make the entire manifold and turbocharger glow red or orange. If this is the case, ignition timing should be advanced to reduce exhaust temperatures. If the knock limit is reached before exhaust temperatures are acceptable, more fuel enrichment is required. Try incrementing lambda down by approximately 0.02 and run the test again with the increased timing. It is acceptable to see some amount of color change even with a cast iron manifold and proper tuning under extended loading, but it must not be left unchecked. Most OEMs limit exhaust temperatures to approximately 1,600 degrees F (870 degrees C) to aid durability. Excessive heat can quickly damage valves, catalysts, turbine wheels, manifolds, and surrounding engine compartment items.

Nitrous Oxide

The final form of forced induction to be discussed here is the use of nitrous oxide gas injection. Nitrous oxide (N_2O, "nitrous") is a molecule that is 33% oxygen, roughly double that of the air we breathe. Nitrous oxide is stored in high-pressure cylinders as a liquid. When released through a valve to ambient pressures, its temperature drops to well below

Cast iron glows orange at a temperature of approximately 1,300°F (~700°C). Most OEM applications target component temperatures below 1,600°F and are calibrated accordingly. This application has heat shields to protect surrounding components.

The components of a nitrous oxide kit are fairly simple. Most kits contain a tank, high-pressure lines, activation solenoids, metering jets, and the necessary wiring.

freezing. The result is a double benefit of both higher chemical oxygen content and lower charge temperatures. The use of nitrous oxide in internal combustion engines was pioneered by German fighter pilots who needed extra help with climbing power during dogfights. By injecting the gas into the engine with the appropriate fuel, power was substantially increased. Nitrous oxide breaks down inside the combustion chamber to allow more raw oxygen molecules to interact with more fuel, creating more heat and cylinder pressure. This increase in pressure and heat is quite large, so it can usually be applied only for a brief period. Nitrous oxide works incredibly well for short-duration events such as drag racing or sprints where heat buildup is not much of a concern.

In modern performance systems, nitrous oxide is stored remotely in a high-pressure cylinder that feeds electronic solenoids metering flow to the engine. The nitrous oxide gas is only injected when extra power is needed at WOT. Extra fuel to accommodate

Nitrous oxide gas is stored in high-pressure tanks usually mounted in the trunk of the vehicle. To ensure proper performance, a heater is sometimes used to maintain a consistent feed pressure to the solenoids as the bottle is drained. (Nate Tovey)

In order to retard the ignition timing only when the nitrous system is active, many racers use an outboard control like the one seen here. The dial allows the user to adjust the amount of spark retard to be applied. (Nate Tovey)

the added oxygen is added through either a secondary feed near the nitrous oxide injection point or by artificially increasing flow through existing injectors. Additional injector flow may come from an increase in fuel-rail pressure or pulsewidth. Care should be taken in the placement of the nitrous oxide injection point. It is possible to freeze MAF and IAT sensors if the injection nozzle is too close or aimed directly at the sensor.

No change to any engine system or performance is necessary when the system is not actively flowing. Calibration at idle and part load are completely unaffected by its presence. Prior to the addition of nitrous oxide, the engine should be completely calibrated at WOT naturally aspirated. This determines future fuel and spark needs when the nitrous oxide is added. If the engine does not operate properly in naturally aspirated trim, it can hardly be expected that performance will improve with the addition of nitrous oxide.

The actual nitrous oxide system is triggered electronically via a relay. There can be a series of switches daisy-chained to limit activation to the desired time. All systems employ some form of arming switch to enable. A WOT switch is also mandatory to prevent accidental engagement at part load. Because nitrous oxide is so effective at increasing engine torque, it should not be used any time torque is being limited by throttling. Introducing nitrous oxide at part throttle causes serious damage because the calibration has the engine operating near $\lambda = 1$ and MBT. The addition of massive amounts of extra oxygen and fuel under these conditions can send cylinder pressures well beyond the mechanical limits of the engine components, even before knock occurs.

The metering of nitrous oxide delivery is regulated by flow through a fixed orifice (jet) at the instantaneous pressure ratio. Its volumetric flow is not linked to engine speed or load. This means that conventional nitrous oxide delivery systems provide an almost immediate increase in oxygen supply and the resulting torque increase. Mechanical limits of the engine must be considered again when choosing a point at which to begin the addition of nitrous oxide at WOT. It is not advisable to attempt to triple the torque output just off idle, as the resulting strain can cause internal engine component failure. The engine must be stabilized under WOT conditions before nitrous oxide is added. This typically means achieving a minimum engine speed of roughly 2,500 to 3,000 rpm. At this point, the onset of torque is smoother and the likelihood of component failure is much lower. For vehicles equipped with automatic transmissions, it is common to set the triggering point for nitrous oxide activation above the stall speed of the torque converter to better control engine loading. Many systems incorporate an engine speed sensitive switch in series with the master arming switch to ensure proper delay.

The addition of the extra oxygen to the intake charge represents a very real increase in load and volumetric efficiency, whether the PCM registers it or not. As a result, spark and fuel needs are similar to an equivalent supercharged engine seeing instant full boost. The difference with nitrous oxide is that there is no real ramping up of load, but rather a step change.

Target lambda varies depending on how much nitrous oxide is to be used. The chemical makeup of nitrous oxide is different from the air, therefore its stoichiometry changes even when burned with the same fuel. The stoichiometric air/fuel ratio for nitrous oxide and gasoline is approximately 7:1, so the target air/fuel ratio should be adjusted accordingly to avoid a lean condition. Since airflow is directly proportional to power potential, target ratios can be calculated based upon what percent of power comes from air versus nitrous oxide. Each fraction of power production is multiplied by its own target ratio. The two are added to find the new target ratio. If a 300-hp engine is found to operate best at 12.7 a/f and 100 hp of nitrous oxide is to be added:

New Target A/F ≈

$$\left[\left(\frac{300}{400}\right) \times (12.7)\right] + \left[\left(\frac{100}{400}\right) \times (7)\right]$$

$$\approx 11.3 \text{ A/F ratio}$$

Base engine fuel maps should not be adjusted at this point. Doing so is only detrimental to normally aspirated engine performance. Instead, nitrous-specific fuel additions only should be adjusted. This typically means adjusting jet size much like on a carbureted engine until an acceptable ratio is reached at all speeds. Some more advanced aftermarket EFI systems allow for nitrous specific fuel adders that can increase pulsewidth for additional delivery. While this method gives a good starting point, the actual ratio may need to be adjusted if combus-

tion or exhaust temperatures become unacceptable.

Ignition lead must be retarded slightly to compensate for the higher combustion temperature and faster burn rates. Most nitrous oxide kit manufacturers agree that a retardation of 2 degrees for every 50 hp dosage increase is a good starting point. For example, a "150-hp" nitrous oxide kit would require 6 degrees of ignition retard at the onset of the tuning process. If lambda is correct, timing can be carefully added to increase engine output until the knock limit is found. This testing should be done with bottle pressure near the intended maximum usable range to ensure that cylinder pressures are near peak during testing. Advancing timing during testing with bottle pressure below target maximum may lead to detonation when the vehicle is run later with a warmer or full bottle.

Much like with the positive-displacement superchargers, timing ends up being lower near peak torque and can increase as engine speed rises. The increase in engine speed and natural airflow due to pumping has the effect of decreasing the percentage of nitrous influence on the power production at higher speeds. This means that the majority of ignition retardation needs to be near system activation and torque peak. Just like the naturally aspirated map, the engine benefits from a slight increase in timing to compensate for increasing speed. Because of the tremendous temperature drop associated with the evaporation of nitrous oxide in the intake charge, ambient temperature compensation does not need to be as aggressive as with a supercharger. In high humidity, nitrous oxide equipped cars can

operate with slightly more timing than their supercharged counterparts due to the latent heat of vaporization of the water vapor in the intake charge. The water molecules absorb some of the combustion heat and energy during the cycle, lowering burn temperatures.

Nitrous oxide can be used in small amounts as a cooling agent for supercharged engines as well. Small dosages (typically less than 50 hp) can substantially drop intake charge temperatures on otherwise heat-soaked charges. These charge cooling doses should be used sparingly, as they add tremendous amounts of cylinder pressure in short order. Fueling should be added accordingly, if not almost excessively. Timing should only be added in very small increments to prevent accidentally venturing into detonation and breaking components without warning. The line between best power and detonation becomes very thin when multiple forms of forced induction are being applied simultaneously.

Some turbocharged race applications use nitrous oxide to instantaneously increase torque during the spooling period. This is only done on small engines with blatantly oversized turbochargers attempting to make large amounts of power. Again, dosages are usually relatively small (usually less than 100 hp), and taper or shut completely off by the point at which full boost is achieved. The nitrous oxide provides for greater mass flow through the engine and exhaust, effectively spooling the turbo like a larger displacement engine at the same speed. To this end, timing is again retarded significantly and excess fuel enrichment is used to maintain combustion stability.

CONCLUSION

If you've made it this far, you should have a good understanding of the inner workings of most engines. Calibration is merely an exercise in giving the engine what it wants at any given time. The entire tuning process helps the PCM predict engine needs based on the inputs it sees. When done correctly, the driver is unaware that the engine's needs are changing, as the output is smooth and responsive to his or her commands. The more precise the calibrator is, the less the driver notices. OEM engineers spend several years sweating the details on their calibrations before releasing a new vehicle or engine to the public, but the procedure is the same for any successful endeavor.

It shouldn't matter what brand of car, engine configuration, or hardware combination is being calibrated.

A logical approach to calibration can often save hours of dyno time and aggravation on the street. When done correctly, the results of a good tune are both good drivability and excellent power. (Nate Tovey)

Good calibration leads directly to driver enjoyment. This car evidently has a good tune, as the driver appears to be enjoying his newfound power a little too much. (Nate Tovey)

Engines all operate on the same principles and laws of physics, thermodynamics, and combustion. Some of the names for components and strategies may change between manufacturers, but the intent remains constant. Any way you look at it, the PCM must process inputs, calculate a model of actual operation, and command outputs to control the physical engine. At the end of the day, the processor on the board of the ECU has no idea whether the fender has a Honda or BMW badge. The calibration process is the same. To reiterate:

- Start with as much information about the engine as possible.
- Input exact component transfer functions into the PCM whenever possible.
- Start with a known good calibration if possible.
- Model fuel delivery as precisely as possible.
- Model engine airflow as precisely as possible.
- Map part-load operation.
- Find a stable idle condition.
- Determine WOT airflow and fuel needs.
- Determine WOT timing tolerance.
- Smooth the transition areas.
- Check corrective values.
- Calibrate transients.
- Test drive.

Following this procedure goes a long way toward preventing headaches. It also allows the tuner to truly perform as a calibrator with precise results and superior engine performance.

The author (left) verifies some data during one of his EFI training class demonstrations on a chassis dynamometer. Students ranged from experienced shop owners to first-time DIY tuners. (Nate Tovey)

Continuing education can help any experienced tuner further sharpen his skills. Students from across the country are gathered at this class taught by the author at Mustang Dynamometer. (Nate Tovey)

FORD TUNING

The Ford Mustang is arguably the most popular vehicle for the do-it-yourself tuner. Ford refers to its PCM as an EEC, or electronic engine control. Starting with the 1988 California specification, and 1989 50-state versions, the Mustang has been equipped with a mass air, sequential EFI system. Other truck and passenger car applications soon followed suit. Various iterations of the EEC have been released with increasing clock speed and capabilities. The EEC-IV systems used on the 1989–1995 OBD-I vehicles were extremely well received by the aftermarket community for their ease of programming and relatively simple control strategy. The mass air based system allowed a large amount of flexibility and ability to adapt at elevated power levels. With a scaled MAF and larger injectors, it was not uncommon to see unmodified EECs supporting over 600 hp. Although drivability was not ideal, the engine operation was acceptable to the performance enthusiast who valued quarter-mile ETs over street manners. The advent of custom tuning software for these EECs allowed experienced calibrators the opportunity to deliver tremendous horsepower and excellent street manners with the stock EEC hardware. Programmable "chips" were designed as modules that could be plugged into the J3 service port opposite the wiring harness on the EEC to hold new operational code for the EEC. When the EEC is booted, the J3 port is examined for data. If an aftermarket module with a valid program is plugged in, the EEC reads the file on it and operates entirely based upon data stored on the new module. If no module is present, the EEC reverts to the program hard coded onto the stock board from the factory.

The 1996 Mustang brought in the era of modular engines and OBD-II controls with the EEC-V. Software companies responded with programming tools that offered the same

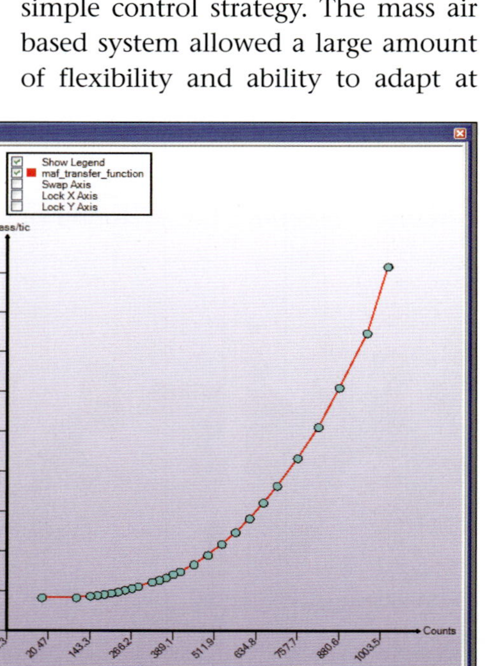

Counts	#mass/tic	Voltage	lbs/min
1023.9	0.0000032731	5.0044	49.0965
1023.0	0.0000032731	5.0000	49.0965
962.0	0.0000026081	4.7019	39.1215
880.0	0.0000020740	4.3011	31.1100
818.0	0.0000016889	3.9980	25.3335
757.0	0.0000013848	3.6999	20.7720
696.0	0.0000011138	3.4018	16.7070
655.0	0.0000009494	3.2014	14.2410
614.0	0.0000007972	3.0010	11.9580
573.0	0.0000006603	2.8006	9.9045
532.0	0.0000005462	2.6002	8.1930
491.0	0.0000004381	2.3998	6.5715
450.0	0.0000003464	2.1994	5.1960
409.0	0.0000002756	1.9990	4.1340
389.0	0.0000002509	1.9013	3.7635
368.0	0.0000002174	1.7986	3.2610
348.0	0.0000001941	1.7009	2.9115
327.0	0.0000001746	1.5982	2.6190
286.0	0.0000001313	1.3978	1.9695
266.0	0.0000001122	1.3001	1.6830
245.0	0.0000000959	1.1975	1.4385
225.0	0.0000000782	1.0997	1.1730
205.0	0.0000000642	1.0020	0.9630
184.0	0.0000000530	0.8993	0.7950
164.0	0.0000000433	0.8016	0.6495
143.0	0.0000000363	0.6989	0.5445
102.0	0.0000000195	0.4985	0.2925
0.0	0.0000000195	0.0000	0.2925
0.0	0.0000000195	0.0000	0.2925

This screen shot from the SCT Advantage software shows the MAF transfer function of a 2001 Mustang Cobra. The EEC actually processes in A/D (analog to digital) counts, but SCT has done the math to show the calibrator actual voltage in the table as well for comparison.

The SCT tuning package puts everything at your fingertips to modify and datalog almost any modern Ford vehicle. (Nate Tovey)

flexibility on the newer EEC-V with the benefit of the finer control strategy from Ford available to the aftermarket tuner as well. The EEC-V family retained the popular J3 port available for aftermarket chip interface, but could now also be flash programmed via the OBD port under the dash. The 2002 EEC-V added another twist with an internal limit to the maximum calculated MAF rate that was reduced to a little over 1,700 kg/hr. Since many supercharged applications can exceed this, scaling of the MAF/injector size/engine displacement became necessary at high power levels on the later versions.

In 2005, Mustang moved to a new family of engine controller known internally as the "Oak" family, with variants spreading across the entire product line. This new processor remains mass air based, but now includes torque-based ETC control. Modifications are still possible, but the torque-based ETC strategy poses a significant challenge to inexperienced calibrators who wish to drastically increase engine power. The trusty J3 port on this processor is gone as well. The only way to change calibration data is by flash programming. To further complicate things, the new CAN

(controller area network) protocol is used for communication rather than the older KWP (key word protocol) as seen in earlier OBD-II EECs. This limits access to scan tools and programmers that have been updated with the newer CAN communication strategy.

Ford EECs convert most voltage inputs from sensors (MAF, TPS, etc.) into an A/D count before actual processing. Some software packages normalize this to only show voltage values in the editor; others leave it as counts. The actual scaling depends upon the clock speed of the processor being used, but it always works out such that 0 to 5 v reference signals become 0 to 1,023 A/D counts. Clock speed of the processor determines the sampling rate, so time-based sensor inputs such as MAF must be corrected before processing. Many actual raw MAF transfer functions are listed "pound mass per clock tick" rather than "pounds per hour," so this can lead to confusion when attempting to copy a MAF transfer function from one model into another if the normalization is not correct.

The Ford EEC-IV/V uses a Hitachi-manufactured MAF sensor with a heated wire element. This design provides a temperature-compensated

mass flow measurement directly to the processor. The post in the center of the MAF helps to reduce the effect of standing waves in the inlet tract on actual MAF measurement. Actual output of this MAF sensor ranges from 0 to 5 v, with a 0 lb/hr= 0 v intercept. The EEC only recognizes a maximum input value of 5 v, even though most sensors continue to increase output voltage with respect to flow all the way up to battery voltage. To prevent "pegging" the EEC's MAF input, the range of the sensor being used should be selected to match the intended maximum engine airflow rate. Many stock calibrations also limit the maximum recognized MAF input to about 4.7 v to account for build tolerances and voltage drift. Since the slope of the MAF transfer function is so steep in this range, the addition of ~0.2 v worth of range (4.9 v max) can usually safely allow for more measurement capacity with the same hardware.

The later 2002 and up EEC-V controllers employ another unique variable Ford calls "max air charge multiplier" that must be changed when adding a supercharger. This variable limits the effective maximum load calculations. Most naturally aspirated applications have this set to 0.9. Changing this to 1.9 (the upper limit in the software) allows the EEC to properly compute loads well beyond 100%. This is also sometimes accompanied by the variable "anticipated air charge multiplier" which should be set slightly lower at approximately 1 8.

Over the years, many of the OE parameters have become accessible to the public. Companies like Superchips Custom Tuning and Diablosport offer excellent calibration tools that display all the essential calibration parameters in a friendly Windows-based environment. These software and hardware

A cold air kit installed in front of the factory MAF alters the way air flows into the sensor. The new MAF output versus actual airflow must be entered into the EEC for proper fuel, load, and spark calculations. (Nate Tovey)

One of the most popular upgrades for the 5.0L Mustang crowd is the Vortech supercharger. When recalibrated with the proper MAF and fuel system, normal driving can be as good as stock with almost double the power.

packages supply everything needed to start from the OEM calibration and change any scalars, functions, or tables needed to accommodate the performance enthusiast's needs. The appropriate starting file is selected by noting the "catch code" marked on both the EEC and the inside of the passenger doorjamb. Once this file is opened up in the editor, changes can be made and saved as a new tuning file. This new tuning file can then be either flashed to a chip to be plugged into the external J3 port of the EEC (the opening with an exposed edge connector opposite of the wiring harness) or stored in a handheld programmer for direct EEC reflashing via the OBD-II port under the dash. Let's look at a couple of specific examples:

Example 1:

A new MAF sensor is installed along with a cold air kit on a 1999 Mustang GT automatic. "CHH2" base file is loaded and saved with a new tuning file name to avoid overwriting the original. Under the "functions" group, there is a listing for MAF transfer function. If the new MAF has a known output, this can be copied into the transfer function as a starting point. SCT offers a handy list of almost every MAF available for Mustangs that can be pasted into the editor in seconds. At this point, the tune should be relatively close, at least enough to start and drive the vehicle. As long as the error between actual mass flow and MAF sensor output is within about 10%, the closed loop learning of the EEC fills in the gaps enough for most conditions. However, the proper recalibration should include modeling of the airflow to accommodate the new actual transfer function of the MAF sensor and cold air kit.

Since the inlet plumbing upstream of the MAF skews its output, the proper method is to set target λ = 1, force open loop, and map each MAF voltage (or A/D count) break point for actual airflow under part load. Repeat for WOT at a richer lambda to build the upper portion of the MAF transfer function and return to closed loop operation. This ensures proper load calculation and fuel delivery under all conditions. Once this is known, a new WOT lambda can be set in the base fuel table (at 70 to 90% load) along with a new borderline spark advance value at WOT. Ford calibrates the Mustang GT for 87-octane fuel, so there should be room for a couple more degrees of advance with 91-octane or higher. The result should be an increase of about 15 hp over stock, depending on how aggressive one gets with the spark advance tuning.

Example 2:

A 9 psi maximum centrifugal supercharger, MAF, and injectors are added to a 1992 Mustang GT manual. "A9L" base file is loaded and saved with a new tuning file name to avoid overwriting the original. Again, starting with the OEM file for this vehicle saves a lot of time sorting out much of the background functions. We focus on the parameters that have been directly affected by the addition of the supercharger and fuel system.

First the injector constants must be adjusted to accommodate the new hardware. If this example uses Ford/Bosch 42 lbs/hr (green top) injectors, the appropriate values can be

Figure A-1 *The voltage compensation curve of the 42 lbs/hr injector is inserted to replace the original values. The SCT software highlights values higher than the stock reference in red, and lower values in blue.*

copied from their original application (1999 F150 Lightning). This means a low slope of 50.0, a high slope of 42.3, break point of $2.5e^{-5}$, and the corresponding voltage compensation curve (as seen in **Figure A-1**). These changes can be easily made by either loading the SCT value file for "42-lb injectors" or selecting "Bosch 42 lbs/hr" injectors in the Diablosport tuning wizard. After a quick check to the scalars tabs to confirm the new values are in place, we can proceed. If a different tuning software package is used, these values must all be entered manually. Remember, properly modeling the fuel delivery is absolutely critical to the accuracy of airflow modeling to be done later.

Idle speed should be increased slightly to prevent possible stalling. The factory setting of 672 rpm may not be able to support the drag of the supercharger. Changing the target speed to about 750 rpm goes a long way toward providing a stable idle. Since the car is likely running on 93-octane fuel, the ignition lead can also be increased at light load to offset the slower burn rate. The values in the base ignition table can be increased for all loads below about 80% by about 3 to 4 degrees, or this particular EEC allows us to simply enter 4 degrees as a scalar under "closed throttle spark adder" and "part throttle spark adder" for quicker results.

The base timing table only maps loads up to about 90%, so this table can be left essentially unchanged for much of its range. The top row of this table should have its high RPM values reduced slightly to smooth the transition into boost. The 1989–'93 EEC provides a specific table for WOT spark advance, so this is reshaped to reduce spark advance in areas where the cylinder pressure is expected to be higher. On later EECs, the base table is rescaled to include loads in excess of 100% and the appropriate cells are changed to optimize timing under boost.

With the centrifugal supercharger, WOT ignition timing below 2,500 rpm can be left unaltered. As boost is expected to build near 3,500 rpm, timing is reduced from about 22 to 20 degrees to allow for the slightly faster burn rate as density increases. Near 6,000 rpm, manifold pressure is highest along with intake temperatures due to compression from the supercharger. Timing is changed from the factory setting of 26 degrees down to about 19 degrees. Later dynamometer testing reveals exactly how much ignition advance can be added back in before detonation, but this should be done only after delivered lambda matches the desired safe target ratio.

Next, a good starting point for the MAF transfer function must be chosen. If using a "scaled" MAF for this application such as a Pro-M 80 mm piece that is "calibrated for 42 lbs/hr injectors," a

Continued on page 110

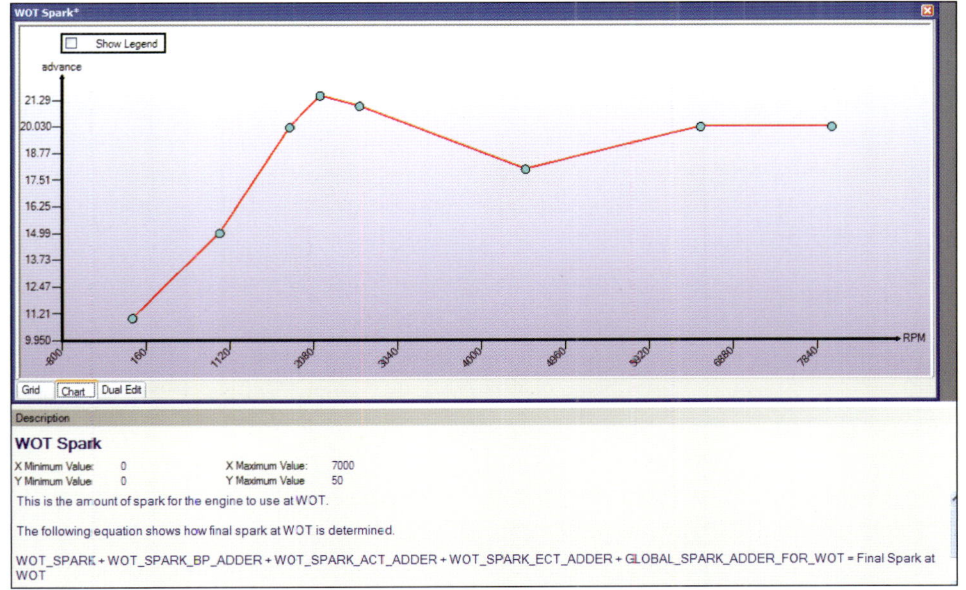

A typical WOT timing curve for a centrifugally supercharged 5.0L engine is shown using the SCT Advantage software.

SCT Advantage 3.0 in action

It's time to get started doing some real calibration work. If the test subject is a Ford vehicle made after 1989, chances are really good that you'll want to consider using a tuning package like this one from SCT. Advantage is an orderly layout for the calibrator that separates tuning parameters into common groups, regardless of where in the actual code the variable is hidden. The result is a clean listing with all of the MAF sensor values under one heading, the injector values under another, and so on. This makes checking your work that much easier when hunting for related variables.

Upon startup, the program prompts the user for the EEC's box code (**Figure A-2**). Upon entering the desired code (or the actual EEC strategy label if you prefer), advantage provides a listing of all available parameters on the left side of the screen (**Figure A-3**). The values shown in each table are the stock values for original OEM calibration code. Since the object is to recalibrate the EEC to accommodate a new combination of parts, one can begin to see where changes must be made.

For example, if the fuel injectors on a Mustang GT are changed from 19 lbs/hr to 30 lbs/hr pieces, we know the fuel-delivery model must be adjusted accordingly. Not only will the static flow rate of the injector require adjustment, but the voltage compensation and starting pulsewidths will as well. Normally, this is the point where we would look up the appropriate values from our reference materials and enter each new point by hand. The

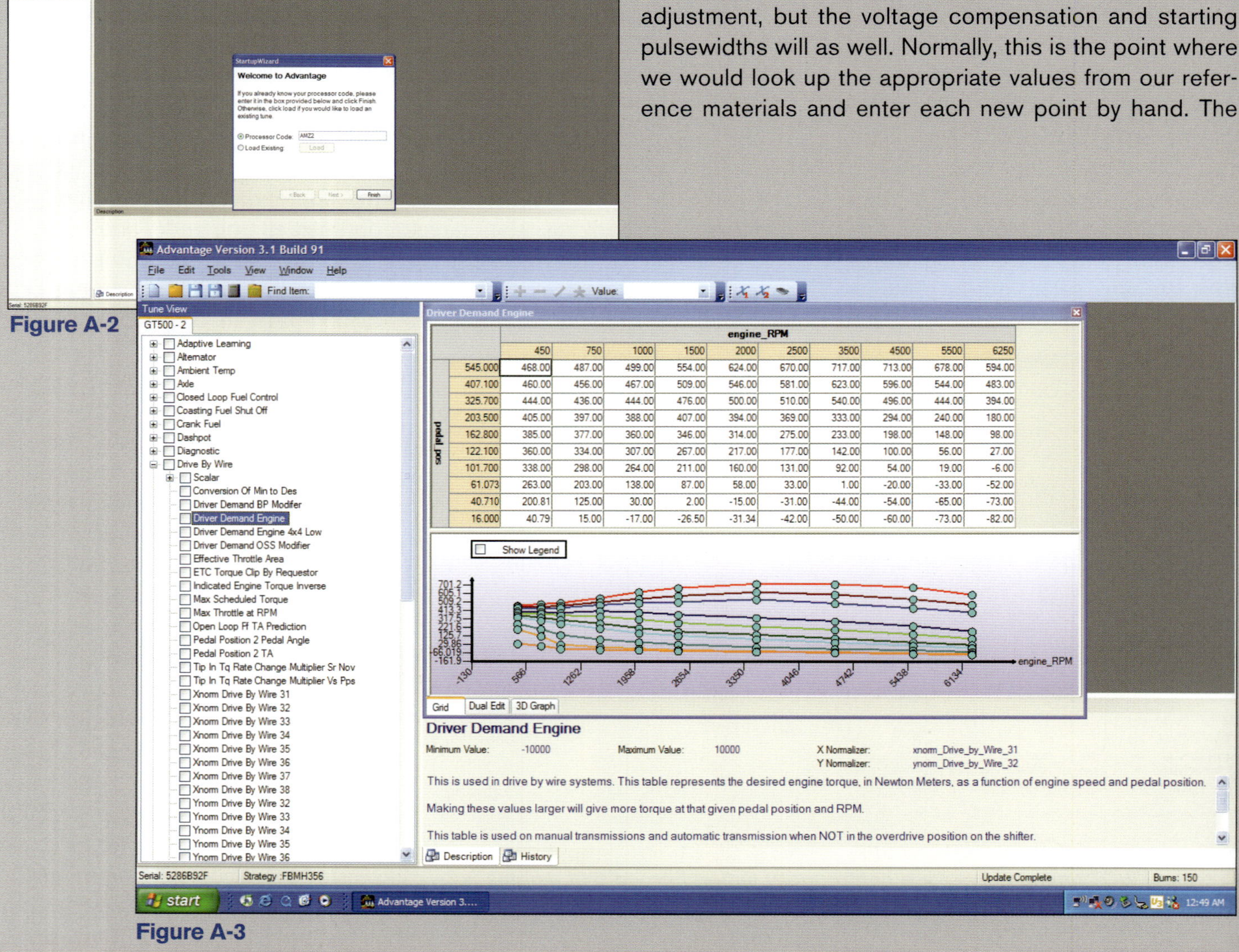

Figure A-2

Figure A-3

pedal_pos				engine_RPM						
	450	750	1000	1500	2000	2500	3500	4500	5500	6250
545.000	468.00	487.00	499.00	554.00	624.00	670.00	717.00	713.00	678.00	594.00
407.100	460.00	456.00	467.00	509.00	546.00	581.00	623.00	596.00	544.00	483.00
325.700	444.00	436.00	444.00	476.00	500.00	510.00	540.00	496.00	444.00	394.00
203.500	405.00	397.00	388.00	407.00	394.00	369.00	333.00	294.00	240.00	180.00
162.800	385.00	377.00	360.00	346.00	314.00	275.00	233.00	198.00	148.00	98.00
122.100	360.00	334.00	307.00	267.00	217.00	177.00	142.00	100.00	56.00	27.00
101.700	338.00	298.00	264.00	211.00	160.00	131.00	92.00	54.00	19.00	-6.00
61.073	263.00	203.00	138.00	87.00	58.00	33.00	1.00	-20.00	-33.00	-52.00
40.710	200.81	125.00	30.00	2.00	-15.00	-31.00	-44.00	-54.00	-65.00	-73.00
16.000	40.79	15.00	-17.00	-26.50	-31.34	-42.00	-50.00	-60.00	-73.00	-82.00

folks at SCT have sped this process up by supplying some shortcuts in the form of their "value files." These value files contain blocks of data that can be copied onto the working program within Advantage to update files quickly. In our test case, the "Load All Values" option is selected and we choose the 30 lbs/hr injector value file from the listing.

Once this value file is applied, we can confirm the update by opening the "Fuel Injector" heading on the left. The "scalars" tab opens a window on the right showing the current values (**Figure A-4**). If these values are correct, we are clear to proceed with the next calibration update. It's a good idea to always check up on these changes. Occasionally, a software glitch may prevent all of the variables from being properly changed for a specific application. Double-checking the updates saves headaches when depending on these values for airflow model correction later.

The process is repeated to update the known good base (OEM file) with the known changes (value files) for all appropriate changes. SCT has value files written that update for MAF sensors, injectors, superchargers, gear ratio changes, and basic N/A power increases. It is strongly recommended that the calibrator at least review the individual changes being made by these value files so that he understands how the EEC is asking the engine to perform before proceeding.

At this point, the file should be close enough to start and run the engine briefly. To check this, the file is saved with a unique name and flashed to either a chip or handheld programming tool by clicking a quick link icon near the top of the screen. Clicking brings up the programming dialog to confirm file name and desired position selection. The "Eliminator" chips and "XCalibrator2" can both hold multiple programs for the same vehicle. Once saved to the appropriate media, the tune can be applied to the car either by plugging the chip into the J3 port or flash programming via an OBD connection.

After all basic changes have been made using value files, we proceed to preparation for actual calibration. The engine is started and we begin sampling wideband lambda readings to confirm the accuracy of the airflow model. If the calibration has been locked into open loop, the correction process is as easy as recording delivered lambda against MAF output. A simple table of MAF values with room to write delivered lambda readings can make quick work of adjusting the MAF transfer function to correct the airflow model. Remember, for areas where target $\lambda = 1$, delivered lambda becomes the correction factor for the MAF transfer function.

Spark values are pretty close to optimal when using one of the SCT value files. Using a chassis dynamometer as described previously, work up through as many speed-load points as possible at $\lambda = 1$ to adjust the MAF curve. You should be able to get about halfway through the curve at $\lambda = 1$. Stop and apply the noted corrections to the file and extrapolate upward if you see a trend. Upload the tune to the vehicle and return to testing. As long as lambda values are closely following the targets, it should be safe to proceed to some WOT pulls to map the remainder of the MAF transfer function.

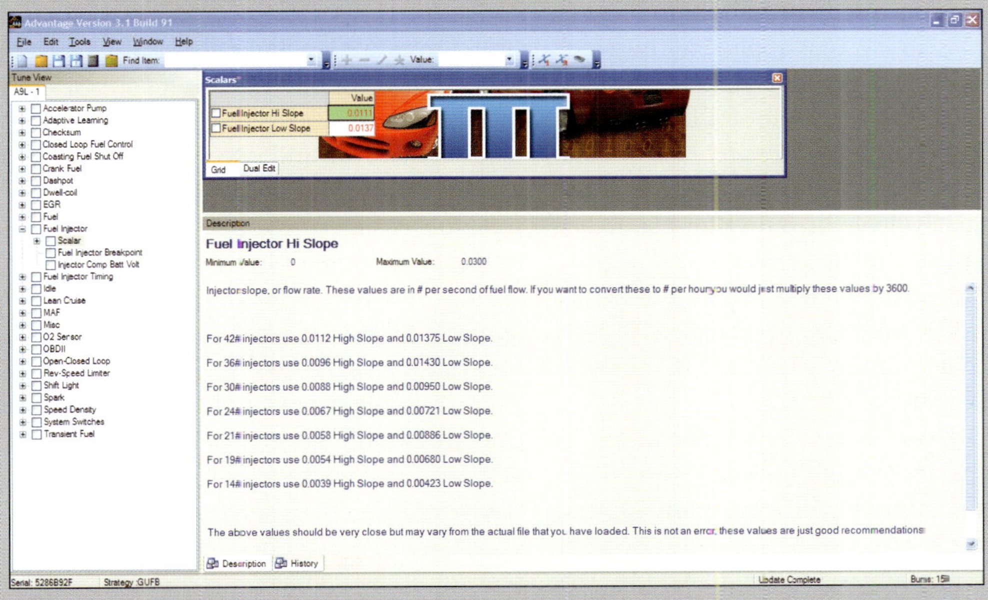

Figure A-4

2002+ EEC-V Tricks

The later 2002-up EEC-V controllers employ another unique variable Ford calls "max air charge multiplier" that must be changed when adding a supercharger. This variable limits the effective maximum load calculations. Most naturally aspirated applications have this set to 0.9. Changing this to 1.9 (the upper limit in the software) allows the EEC to properly compute loads well beyond 100%. This is also sometimes accompanied by the variable "anticipated air charge multiplier," which should be set slightly lower at approximately 1.8.

These newer EECs can sometimes prevent the use of a MAF sensor's exact transfer function. Since they are internally limited to a maximum calculated airflow value of approximately 1,700 kg/hr, some scaling must be done to stay within this limit and allow the EEC to remain in control of fuel mixture at high load. This is done by scaling the MAF sensor transfer function as previously discussed. If the projected actual MAF range is 20% higher than this limit, the curve must be shifted by a multiplier of 0.8 to remain within range of the EEC.

At this time, a few other key changes must be made to ensure proper fuel control. The most important change is the flow rate of the injectors (both high and low slopes). They must be scaled by the same multiplier of 0.8 in this example. The engine displacement scalar is often labeled "displacement of one cylinder," and should be scaled by the same multiplier of 0.8 to maintain proper load calculations.

The scaling process is not quite complete for a Ford at this point. Additional scaling must be applied to both the "intake manifold volume" and "injector break point" scalars to correct transient fueling calculations. Ford uses intake manifold volume as a primary input to the acceleration enrichment logic to predict necessary wall film compensations. Since these are essentially load-based calculations, it is helpful to scale this parameter to improve the precision of the tip-in fueling calculations.

simple multiplier can be applied to the stock transfer function. This multiplier should be the difference in injector sizes. In this case, a multiplier of $42.3/19.6 \approx 2.16$ is applied to the Y-axis (actual mass flow) of the transfer function as a starting point for later calibration. The SCT software has a large number of MAF sensor transfer functions available in the value file library to make this step quicker and slightly more accurate. Regardless, the actual transfer function as installed in this specific vehicle must still be mapped on the dynamometer.

Now open loop operation is forced by setting the closed loop enable temperature to the maximum allowed and the target fuel maps must be reset to ease mapping of the new MAF. The target lambda for loads under 90% is set to 1.0 (14.68 air/fuel), and the vehi-cle is driven on the dynamometer to measure the error in the MAF transfer function. Lambda values are used to correct the MAF curve in the software until errors are less than about 3% for all light load conditions. Target lambda is then set to $\lambda \approx 0.78$ (11.4 air/fuel) for loads above 70% and sweeps are made to map the top end of the MAF transfer function. Again, corrections are applied to the MAF curve in the software until error is less than 3%. Once delivered lambda is equal to target lambda, timing can be advanced slightly under load to find best power or the detonation threshold. Since these EECs lack any form of knock control, it is advisable to stay about 3 degrees below the knock limit for any street-driven car to prevent future damage.

After this mapping has been done, the base fuel table can be restored to a smooth transition from $\lambda = 1$ at idle and cruise to the desired ratio under high load. Closed-loop operation can be reinstated by setting the enable temperature back to stock. If the vehicle has a slight hesitation on tip-in, the acceleration enrichment can be increased to offset the temporary lean condition on transition. This is done by changing the "AE multiplier vs throttle position" curve. This compensates for the mechanical lag in the MAF measurement caused by the longer path between the meter and throttle body. It also compensates for the increased instantaneous flow that comes with positive pressure built up against the outside of the throttle blade from the compressor.

At this point, the vehicle is ready to be test driven to check for any other issues.

GM TUNING

The GENIII (LS1, LS6, LQ9) engine family represents a large step forward in performance and technology for GM. The only things it has in common with its predecessors are bore spacing and manufacturer's label. Although still pushrod activated, it rivals the specific output and emissions of the most advanced OHC competitors with potential for tremendous performance increases.

GM PCMs used in the early TPI systems were a simple MAF based, bank-to-bank system. Early LT1 systems were speed density with sequential control with knock control. The LT1, 3800 V-6, and LS1 are primarily MAF based, sequential injection with knock control. For LT1, spark is calculated in terms of MAP, just like speed density. For LS1 applications, load is calculated as g/cyl.

Engine knock is monitored by a sensor installed into the water jacket. GM knock strategy as a whole is generally good. Because of the accuracy of this system, their engineers were able to deliver production calibrations that ran the engine much closer to the knock threshold resulting in increased power and economy. When calibrating for best power on a modified engine, this knock routine can be a very useful tool. By choosing a slightly aggressive spark curve and allowing the knock circuit sufficient authority (at least 4 degrees of retard possible), the best timing can quickly be found. If the target ignition value is beyond the knock limit, the PCM should detect this knock and the scan tool should display a corresponding retard value that has been applied to alleviate the condition. Recognizing the hysteresis of knock, this whole value does not necessarily need to be removed from the target advance. If the PCM required 4 degrees of retard to eliminate the detected knock, it may only require 2 degrees adjustment to the target value to avoid initially crossing the knock threshold.

This sensor often proves to be sensitive to valvetrain noise that occurs in the same frequency range as true engine knock. For the LT4, a newer, less sensitive sensor was used to reduce the chance of false knock resulting from its slightly noisier roller rockers. Some aftermarket valvetrain combinations may simply prove too noisy for the knock circuit to ignore. In these rare instances, knock can be disabled by setting maximum retard to zero and carefully calibrating the spark tables on a dynamometer. Since the knock circuit is no longer available, a few extra degrees of safety margin should be left to avoid detonation during worst case loading.

GM LT1, 3800, and LS1 applications use similar Delphi MAF sensors.

The GM MAF sensor is also a hot-wire element design. The large bore minimizes pressure drop across the element for maximum power. Output is in frequency instead of voltage like other MAFs. (Nate Tovey)

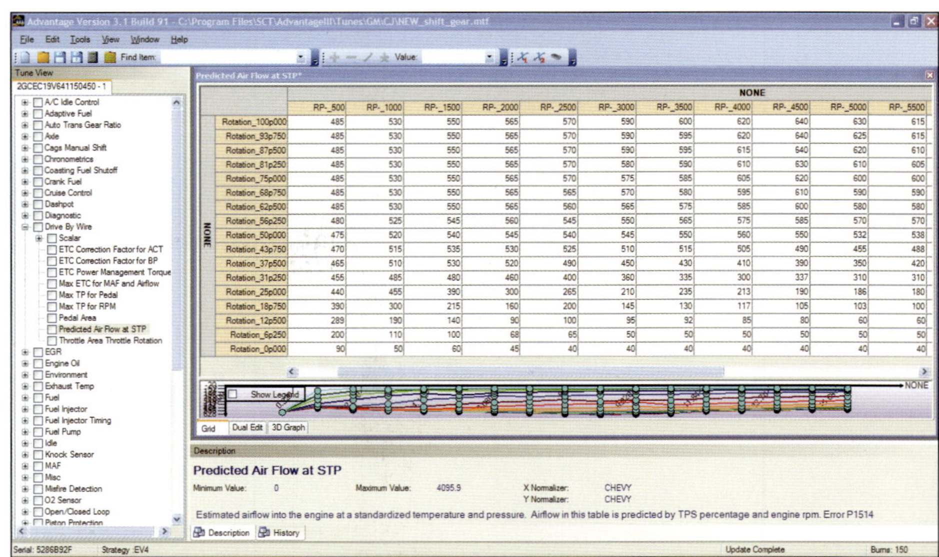

| | NONE | | | | | | | | | | |
	RP-_500	RP-_1000	RP-_1500	RP-_2000	RP-_2500	RP-_3000	RP-_3500	RP-_4000	RP-_4500	RP-_5000	RP-_5500
Rotation_100p000	485	530	550	565	570	590	600	620	640	630	615
Rotation_93p750	485	530	550	565	570	590	595	620	640	625	615
Rotation_87p500	485	530	550	565	570	590	595	615	640	620	610
Rotation_81p250	485	530	550	565	570	580	590	610	630	610	605
Rotation_75p000	485	530	550	565	570	575	585	605	620	600	600
Rotation_68p750	485	530	550	565	565	570	580	595	610	590	590
Rotation_62p500	485	530	550	565	560	565	575	585	600	580	580
Rotation_56p250	480	525	545	560	545	550	565	575	585	570	570
Rotation_50p000	475	520	540	545	540	545	550	560	550	532	538
Rotation_43p750	470	515	535	530	525	510	515	505	490	455	488
Rotation_37p500	465	510	530	520	490	450	430	410	390	350	420
Rotation_31p250	455	485	480	460	400	360	335	300	337	310	310
Rotation_25p000	440	455	390	300	265	210	235	213	190	186	180
Rotation_18p750	390	300	215	160	200	145	130	117	105	103	100
Rotation_12p500	289	190	140	90	100	95	92	85	80	60	60
Rotation_6p250	200	110	100	68	65	50	50	50	50	50	50
Rotation_0p000	90	50	60	45	40	40	40	40	40	40	40

Predicted Air Flow at STP

Minimum Value: 0 Maximum Value: 4095.9 X Normalizer: CHEVY Y Normalizer: CHEVY

Estimated airflow into the engine at a standardized temperature and pressure. Airflow in this table is predicted by TPS percentage and engine rpm. Error P1514

The ETC safety system of the C5 Corvette checks actual current MAF values against this table. If the values in the table are exceeded, a fault is triggered and throttle opening is limited. The table is shown this time in the SCT Advantage software.

These use the familiar hot wire element design, but with a frequency rather than voltage output to the PCM. The range of the factory MAF for these cars is surprisingly wide, making them flexible enough to work even in most supercharged applications. Their design allows them to be installed as either draw-through or blow-through for boosted applications with minimal concerns. The integrated honeycomb on the inlet side serves to straighten the flow in front of the metering elements. Removing this screen does almost nothing to improve flow.

The Delphi sensors also have a "wing" spanning the center of the measurement housing to provide an ideal flow past the actual metering wires. Enthusiasts often eliminate this post, mistaking it for a flow restriction. Doing so does indeed slightly reduce the pressure drop across the MAF sensor, but it also skews the output downward by 7 to 10%. The lower calculated loads both lean out fuel and advance timing, running the risk of premature knock. This lean condition must be adapted by the PCM's closed loop fueling correction strategy before the vehicle drives properly. Even then, actual load calculations will be slightly off during all conditions unless the MAF transfer function is remapped to reflect the new actual output. If a "ported" MAF is to be used, plan on spending some time building a new transfer function before doing much other calibration work.

The GM LS1 PCMs employ a complex active intake manifold model in the form of the "base volumetric efficiency" table. Much like a pure speed density system, this table is a map of predicted engine performance based on speed and MAP. It is used to calculate transient behavior faster than would otherwise be done by reading MAF data. Serious changes to the engine's airflow characteristic (such as a change to a longer duration camshaft) should be accommodated in this table even if the MAF curve is

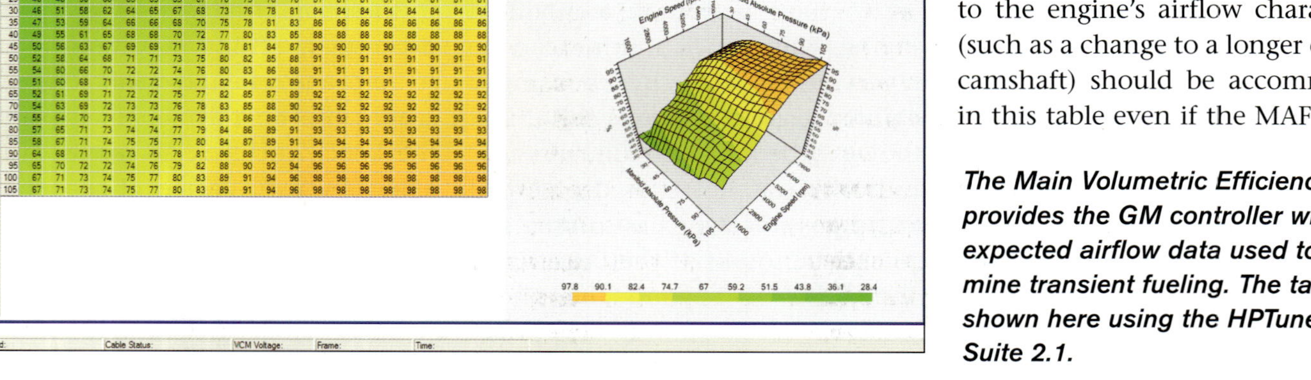

The Main Volumetric Efficiency table provides the GM controller with expected airflow data used to determine transient fueling. The table is shown here using the HPTuners VCM Suite 2.1.

100% accurate. The PCM allows for a great deal of flexibility, but idle and off-idle regions in particular can be sensitive to cam changes. Changing this table to more accurately reflect engine performance goes a long way toward smoothing out startup and low speed behavior. Changing the high load (95 to 105 kPa) portion of this map has the net effect of adjusting acceleration enrichment.

Accurate mapping of this table's data can often be accomplished by simply unplugging the MAF sensor. This triggers a MAF performance fault code and forces the PCM to default to operating based on the values in the base VE table. At this time, the table can be calibrated much the same way as a standard speed density system. It is possible to calibrate just this table and leave the MAF sensor out of the system by turning off the fault codes associated with the MAF sensor. However, better flexibility, weather, and load compensation can usually be had by restoring proper MAF function after the base VE table has been adjusted for the specific engine combination.

With the introduction of the C5 Corvette, GM introduced ETC control on the LS1. This was a pedal follower strategy with a simple safety check table. ETC performance was monitored by checking MAF input against the values in the "g/cyl for engine speed versus throttle rotation" table. If the MAF input exceeds the current value of this table, it is determined that a fault must exist with the mechanical portion of the ETC and a limp-home mode is triggered that limits engine and vehicle speed. The installation of a supercharger onto one of these engines easily exceeds the factory values for this table. To prevent prematurely triggering a limp mode, this table must be adjusted to accommodate the new higher flow rates. Great care should be taken to only increase the cells necessary. This avoids premature safety activation, and even then only by just enough for proper operation. Blindly increasing all areas of this table, or increasing the necessary areas by too much, runs the risk of failing to detect a genuine problem. Any calibrator who ignores the importance of this safety table leaves himself open to a tremendous lawsuit if the vehicle surges out of control causing an accident or injury!

Aftermarket access to most GM engine control modules is readily available today. Packages such as HPTuners, CarPuting's LS1/LT-1 Edit, C.A.T.S., and SCT advantage allow the aftermarket calibrator open access to almost all factory parameters in the PCM. These windows-based tuning packages have the ability to read the stock file from the car, allow the calibrator to edit the files using real engineering units, and reflash the modified file into the vehicle with just a laptop and adaptor cable. Most have access to the same set of GM parameters, with slightly different descriptions or chart layouts. As long as one takes the time to carefully examine the values and layout, almost any of these will get the job done. The user interface and ease of use varies from package to package, so take a look at all of the available options before spending a significant amount of money on either. Let's take a look at a few examples.

Example 1:

A cold air kit, "shorty" headers, and exhaust are installed on a 1995 Camaro Z28 with a manual transmission. First, the stock file is read out from the vehicle and saved. A copy of this

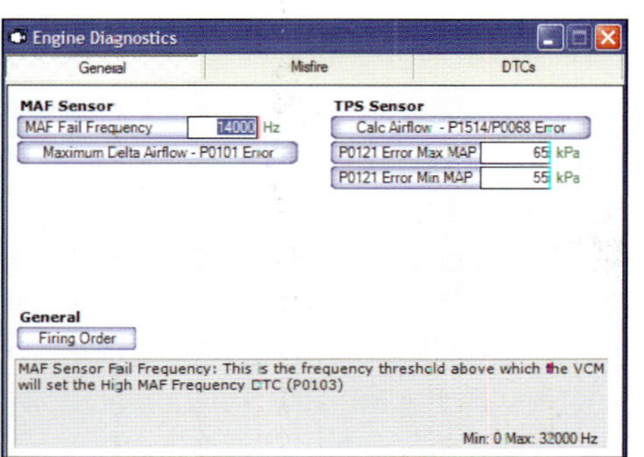

The MAF sensor on the GENIII engine is disabled by setting the MAF Fail Frequency to zero. This forces the computer to ignore the MAF signal and run in speed density full time.

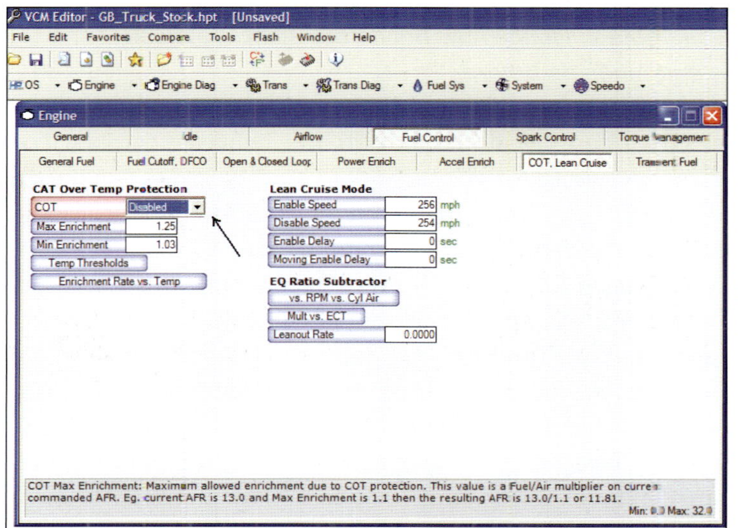

In order to maintain a consistent commanded air-fuel ratio during high-load testing, the Catalyst Overtemp Protection (COP) feature is disabled prior to testing.

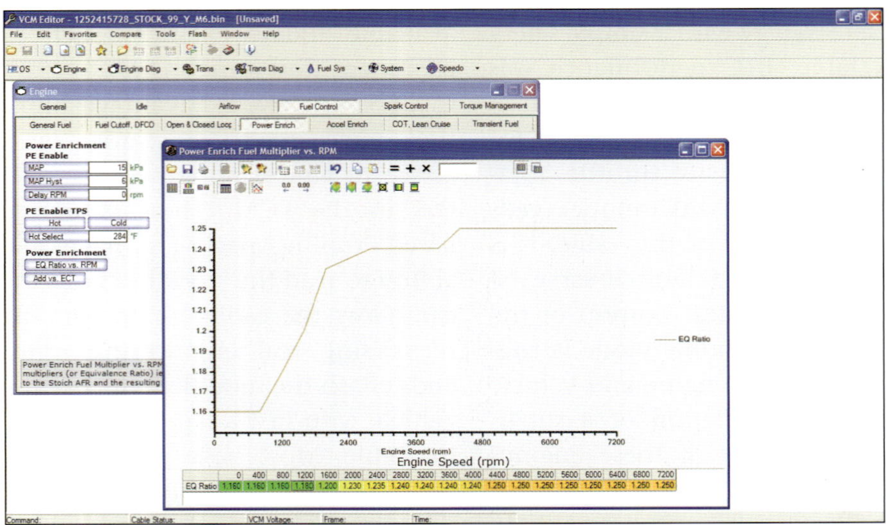

The GENIII PCMs use the "WOT PE vs RPM" curve to dictate an air/fuel ratio under heavy loads. The units are equivalence ratio, the inverse of lambda.

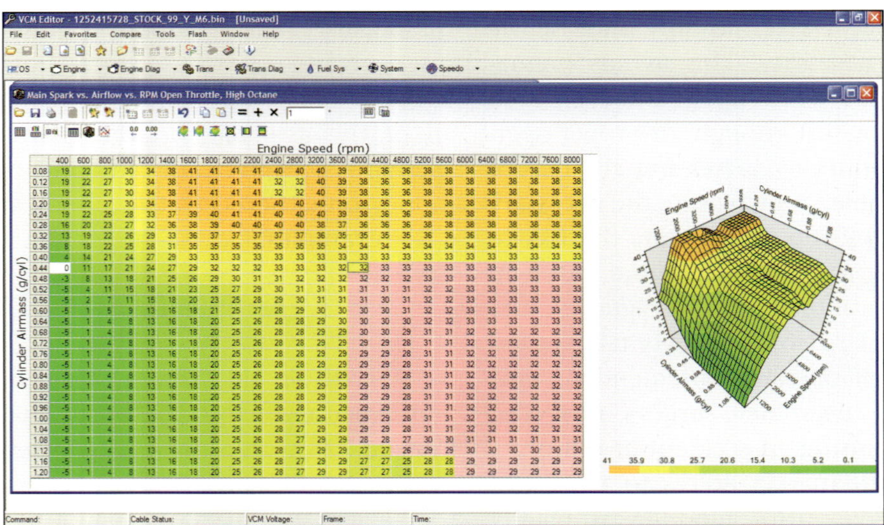

This modified table shows the results of a tuning session with our test vehicle as shown in example 2. Notice how timing does not drastically drop at high speeds and loads as before.

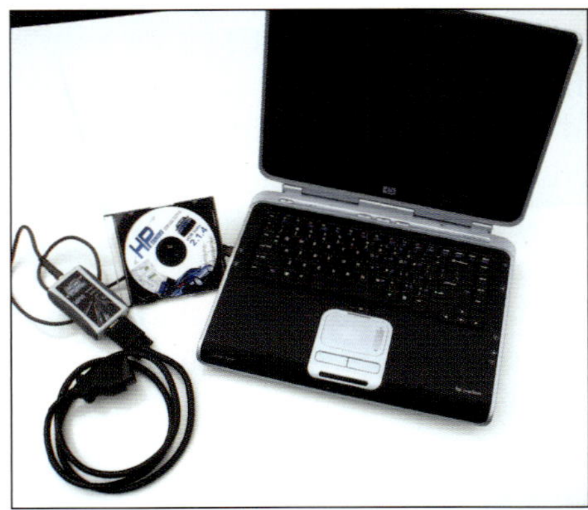

Aftermarket tools like the HPTuners VCM Suite perform double duty as programming and datalogging tools for almost any popular GM application. (Nate Tovey)

file is loaded into the editor and saved with a unique name. Again, we take a moment recognize what physical changes have been made to the vehicle that require calibration changes. In this case, the MAF itself remains stock, but its installation in a new air path may shift its output. The vehicle is driven at a steady speed on the dynamometer, and MAF output frequency is recorded along with fuel trims and delivered lambda. If the delivered lambda corrects back to $\lambda = 1$ at cruise or idle, the fuel trims will show the necessary adjustment to the MAF curve. Typically, this free flowing inlet leads to positive fuel trims (fuel is being added to maintain the desired ratio), so the current MAF frequency is noted along with the average long-term adjustment. This same adjustment is applied to the MAF transfer function to bring the PCM's airflow calculations back into alignment with reality. It is very important that this is done in steady state to avoid the influence of acceleration enrichment. Multiple points are taken along the MAF curve in the cruise areas of the base fuel map. A global trend is usually found in the MAF offset due to the cold air kit. This correction can be applied to the MAF transfer function and extrapolated out to maximum output.

While we have the file open in the editor, let's take a look at a couple other items. Most customers wish to disable the Computer Assisted Gear Selection (CAGS) that forces the one to four upshift at low loads. GM implemented the strategy to force drivers to operate the vehicle in such a manner as to improve fuel economy in the city. This feature can be disabled by setting the minimum activation temperature to a very high value such as 255 degrees F. It can also be defeated by changing the maximum throttle position to enable to zero degrees. Much like the Ford applications, there are performance benefits to triggering fans earlier to improve performance for GM cars as well.

The Chevrolet Camaro SS as delivered with 320 hp in 2002 was king of the ponycar wars. This particular example is equipped with a supercharger, bigger cam, and about double the power. (Nate Tovey)

The fifth-generation Corvette came with the introduction of the LS1 engine family. Output of these factory engines can be more than doubled with the right bolt-ons and calibration. (Nate Tovey)

Looking at the values in the "WOT PE Equivalence Ratio," we can see where GM has calibrated the vehicle to run rather rich to keep catalyst and exhaust gas temperatures down. This table uses Phi (φ, equivalence ratio), which is the inverse of lambda, so $\varphi = 1.25$ equals 25% enrichment, $\lambda = 0.75$, or A/F = 11.0:1. Looking at the factory values here, it's easy to see where significant power increases can be made. By choosing a more reasonable value such as $\varphi = 1.1$, or $\lambda = 0.9$, we see a better power match to this naturally aspirated engine.

Timing can be advanced in the high rpm table too. The installation of a free flowing exhaust reduces back-pressure and cylinder temperatures. This in turn buys us a little greater knock margin. Timing is increased slightly to move closer to MBT at full load. If you go too far here, you are likely to see some activity on the knock retard when datalogging at WOT. Once the vehicle can make a WOT pull to redline without any knock retard activity, the aggressiveness of the factory knock routine can be reduced. This can be done by limiting the "maximum knock versus RPM" table to roughly 4 degrees, lowering the "fast knock gain" table by about 30% above 3,000 rpm, and increasing the "knock recovery" table by about 50% above 3,000 rpm. This still allows the knock routine to remain active, but makes it less intrusive in cases of light knock.

Example 2:

A new camshaft and ported cylinder heads are installed in a 2002 Corvette. The addition of the heads and cam introduces a significant departure from the factory engine performance. As a result, dynamic changes in airflow can be under-predicted by as much as 30%. Correcting this requires fundamental changes to what the PCM expects to see for airflow at a given speed-load point.

As mentioned earlier, the GM dynamic air strategy on LS1 family engines is controlled largely by the "Main VE Table." Any time the PCM detects a change in airflow (not steady state), it references this table for its first attempt at fuel prediction. This can include the rolling or hunting idle that may accompany a large cam at stock idle speeds.

We begin by reading the stock data from the vehicle, saving, copying and renaming our working file. The next step is to set a more realistic idle speed. The LS1 can have as many as four idle speed tables to suit combinations of A/C on-off, and "park/neutral" versus "in gear." All of the appropriate tables should be increased globally to begin.

You can always go back later and slow idle speed to get the "muscle car" idle note and shake if the customer desires later. For cams in the 230 degree (at 0.050" lift) duration range on a stock displacement engine, it would not be unusual to increase warm idle speed to about 950rpm for initial work. This should save some frustration from constant stalling during the tuning process. Remember, you can always go back and lower this after you've calibrated the airpath model correctly.

Fundamentally, the larger cam shifts the power band of the engine upward in speed. As such, low RPM volumetric efficiency usually drops from stock. Somewhere in the midrange there will be an inflection point where the new VE matches the stock values. Above this, the expected VE will be greater than stock. (How else would we be making more power out of the same engine, right?) It is this change in VE versus RPM that must be entered into the "Main VE Table." Otherwise, idle speed changes are met with slight over-fueling (and the appropriate increase in torque/speed), leading to surging. Likewise, higher RPM tip-in events would be under-fueled, leading to hesitation or bucking.

There are several ways to correct the main VE table on an LS1 PCM. Each can be equally effective as long as the calibrator uses a sound approach. The first method would be to disable the MAF sensor completely. The MAF is disabled by setting "MAF Fail Frequency" to zero and disabling error codes P0100/P0101. This forces the PCM into a failsafe mode where the air mass is calculated strictly from the main VE table like a conventional speed density system.

At this point the VE table can be calibrated in steady state, open loop on the dyno using a wideband oxygen sensor or by applying long-term fuel corrections to the values in the main VE table. Generally speaking, the VE changes resulting from a change in cam have a greater dependency upon RPM than MAP. This means that one can usually apply a correction to an entire RPM column (i.e., all load points at 2000rpm would receive the same multiplier) in the main VE table, at least to start. The more diligent one is during this process, the better the transitional behavior will be for the car.

Many DIY enthusiasts and tuners swear by leaving the car in this speed density mode. However, if the MAF is properly calibrated as well, steady state fueling accuracy will also improve. The MAF also gives the PCM the ability to respond to changes in weather conditions more immediately rather than depending upon adaptive strategies from the IAT and MAP sensors. In this case, the MAF transfer can be calibrated as before using steady state and WOT sweep dyno measurements.

The LS1 PCMs also employ a catalyst overtemp protection feature. For initial tuning, this should be disabled to make sure that target WOT lambda remains constant. The WOT target ratio is also set with the "PE versus RPM" table, using equivalence ratio as units. This table should be adjusted to a reasonable target value prior to WOT dyno testing. As before,

errors between the target and delivered air/fuel ratio can be adjusted in the MAF transfer function. If catalyst temperature protection is still desired, it can be reactivated after all WOT mapping has been completed.

Spark tables also require significant changes for this example. The larger duration, and consequently overlap, of the new camshaft results in more natural EGR. This means that there will be an almost global need for increased spark advance, including idle and light cruise conditions. These changes need to be applied to the high and low octane base spark tables as well as idle spark tables for "park/neutral" and "in gear." It is worthwhile to perform a number of checks at constant speed/load on the dyno in the midrange find out just how far off the new MBT timing is from stock. For the 230° camshaft in this example, it would not unusual to need an extra 4 to 5 degrees of advance at part load to reach the new MBT. Making this change has the added benefit of increased efficiency, fuel economy, and throttle response.

WOT spark tuning is performed as normal, with special care given to observing knock retard values when advancing spark near the new torque peak. It will likely be found that a significant amount of spark advance can be added above 5,800 rpm in this case, with continuing progressive increases until redline. Remember, the more usable engine speed/RPM that is added, the more timing will be required to offset the mechanical delay and dropping cylinder pressure. Again, the shape of the torque curve shows where timing should be added to reduce losses in cylinder pressure.

As a final check, the ETC safety strategy should be reviewed. The code 1514 "Maximum Airflow vs. Throttle

Rotation and Engine Speed" table should be checked against current airflow values. Ideally, there should be about 20% safety margin between nominal airflow values and this table. Too much clearance runs the risk of not detecting a genuine ETC fault and too little can have the driver cursing as he constantly hits a false error and the resulting limp mode.

Example 3:

A positive displacement supercharger is added to a 2001 Silverado Z71. Here, we pay special attention to the change in the MAF transfer function due to the supercharger's new inlet system and set the appropriate fuel control values. Since the heads and cam have not changed, it is unlikely that there will need to be much work done to the Main VE Table. Any time the vehicle enters boost, the MAP sensor immediately jumps to the top value for the VE table.

Larger fuel injectors will be the first major change to the hardware. The flow rate of the new injectors should be entered into the "Injector Flow Rate" table. This table uses kPa of vacuum as its input, so be careful when adjusting. "80 kPa" in this table really represents strong vacuum, and "0 kPa" represents atmospheric pressure in the manifold. The slope of this table confirms the decreasing flow with added manifold pressure, as shown earlier in the injector section. (Chapter 5) This table should receive a global shift equivalent to the increase in static injector flow rate. If the new injectors are 90% higher capacity than stock, a multiplier of 1.90 can be applied. If the injector offset versus battery voltage is known, it should be entered into the "Injector Offset vs. Battery Voltage vs. KPA Vac" table. These changes should result in sufficient injector control to continue with MAF signal correction.

Appropriate target air/fuel ratio values should be chosen at this time. All off-boost areas can be set to $\lambda = 1$. Catalyst overtemp protection should be turned off temporarily for tuning. WOT power enrichment should be set conservatively to approximately $\lambda = 0.78$ (PE value of 1.22 in the table) for all RPMs. The trucks employ a WOT PE delay that must be turned off to allow immediate enrichment under boost. During this delay, WOT fueling remains at $\lambda = 1$. The factory value for this is in excess of 5,000 rpm, which would certainly not be a good idea for a supercharged truck! This value should be reduced to a few hundred RPM above idle to avoid lean conditions and knock under boost.

The spark advance tables for this truck were calibrated for 87-octane fuel. When running a supercharger, hopefully we are working with a minimum of 91-octane fuel. Most of the part load tables already have the truck operating at MBT timing during cruise, so these regions will not require much adjustment, if any. Examine the spark advance values in the high-octane table above 0.8g/cyl. This is the region where the truck begins to develop boost. The factory values are very low here to avoid engine damage (knock) during towing even though these load values are almost never reached on a stock vehicle. You may find that the stock values for 87-octane are not that far off from a conservative timing for our supercharged 91+ octane application. The torque curve dictates the shape of the spark curve as long as temperatures remain consistent. Make sure that this table still has a progressive retard of spark with increasing load. That way if the weather changes

later, the new load is detected by the MAF and the appropriate spark advance can still be calculated by the PCM. This has the added benefit of adapting for pulley changes in the future without too much drama.

A key item to check on automatic transmission equipped LS1 PCMs is the "Max Engine Torque" value. Most PCMs have this set to 350 ft-lbs. This should be increased to avoid torque limiting interference with delivered spark timing near the shift points. Increasing this also results in crisper shifts, since more torque is allowed to be delivered to the transmission in the first place. There is a table labeled "Torque Reduction for Upshift" that has a similar effect. This table in used to prevent overpowering the clutches inside of the transmission. Decreasing the values in this table allows increased torque during shifts, but possibly at the cost of transmission durability. 4x4 vehicles also have axle torque limit scalars that should be increased if full torque is desired at all times.

At this point, normal MAF mapping is done as before. Steady state part load at $\lambda = 1$ is performed first, and WOT sweeps follow to map the top end of the curve. Once delivered lambda equals target lambda, attention can be turned to spark advance or leaning out the target PE values. Keep

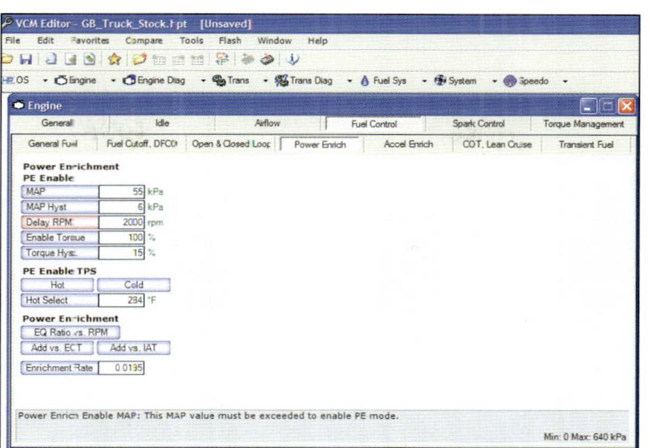

The "PE enrichment delay RPM" is moved from 5,000 rpm downward significantly to provide instant enrichment at low speeds. This supercharger can provide boost very early, so the appropriate fueling must also be added to avoid knock.

Positive displacement superchargers like the Magnuson Radix system shown here are very popular for truck owners since they greatly increase available torque across a wide range. This particular system uses an Eaton Model 112 compressor. (Image courtesy of Magnuson Products)

in mind that the vehicle still has catalysts that must not overheat, so a little richer is better for durability. If catalyst temperatures still get too hot during WOT runs, the COT feature can be reactivated.

The added airflow of the supercharger will almost certainly extend the usable speed range of the engine. After all WOT mapping has been completed, the shift points can be reset to take advantage of the new power band. This is done with both a fixed RPM scalar for each gear and a table showing vehicle speed versus throttle position for upshifts.

STANDALONE *EFI* SYSTEMS

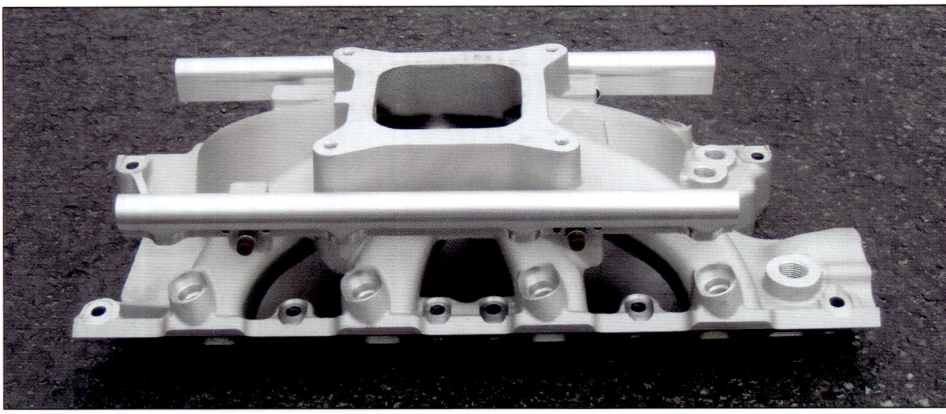

Standalone systems like the F.A.S.T., Accel DFI GENVII, and Big Stuff 3 can be used with specially modified intakes like this one. They are a natural fit for hot rods and restorations where there is little or no existing wiring harness to modify to begin with.

The ever-increasing number of "do-it-yourself" horsepower enthusiasts helped to explode the market for standalone EFI systems. These self-contained systems are designed to work independently from any other vehicle systems. This means they work equally well in a classic restoration project, dedicated racecar, or retrofit to most modern road-going vehicles. Almost all stand-alone aftermarket systems are speed density to reduce the complexity of the MAF and its necessary plumbing. These systems can be installed under the same inlet system as a common four-barrel carburetor or use almost any OEM intake manifold.

Most of these systems allow the calibrator to view data in real time. This means that actual lambda, spark advance, engine speed, load, etc. can all be viewed while making adjustments to the tables on the fly. This reduces the down time associated with taking measurements, changing the file, and reflashing the module before returning to take more measurements. Tuning in real time allows the calibrator to instantly see whether a change was beneficial or not. For steady state calibration, the ability to "arrow up/arrow down" quickly makes airflow modeling and finding MBT incredibly easier.

The complexity of these systems, while intimidating to most beginners, pales in comparison to modern OEM systems. Their calibration is surprisingly simple. Many programs have a tool to estimate a starting VE map or come preloaded with one that is relatively close. If the system is installed and wired properly, one should be able to quickly move on to the business of fine-tuning the base VE map after confirming proper fuel system setup and injector modeling.

Starting with medium load and engine speeds before attempting to idle helps the calibrator identify trends in the engine's VE characteristics that can be extrapolated downward. Taking advantage of the stable nature of the engine at elevated speeds, the VE map is adjusted until delivered lambda equals target lambda at every cell in question. As long as cells exist in enough of the right locations to build an accurate airflow model, fueling should be stable regardless of speed and load. Most engines require more cells in the low engine speed range, since this is where the most drastic changes in actual VE occur.

These necessary shifts can save a lot of time chasing an unstable idle fuel mix that is experiencing misfires from being too far out of range. Likewise, break points necessary to accurately model the airflow can be chosen to improve accuracy. Modeling VE at idle, slightly above, and below gives the PCM the ability to maintain a constant lambda as VE changes, resulting in a more stable idle condition.

Only after steady state, part load, and idle are sufficiently calibrated should WOT tuning be performed. Just like with idle, the trends seen in the base VE map should be extrapolated upward for WOT as well. Again, the VE table is adjusted until actual measured lambda equals the commanded values in the tables.

After all steady state values of the base VE table have been calibrated, we can move on to transient fueling. Acceleration enrichment tables can now be adjusted to provide subjectively good throttle response. Resist the urge to adjust the VE table at this point, since the changes in measured lambda are a result of the wall wetting phenomena under changing conditions.

After the engine has had a chance to cool to ambient conditions, it should be started again to verify calibration of the temperature-compensation curves. In a speed density system, these curves have significant authority on delivered lambda. Many tuners mistakenly recalibrate the base VE table when encountering poor lambda control the day after tuning. If there is a

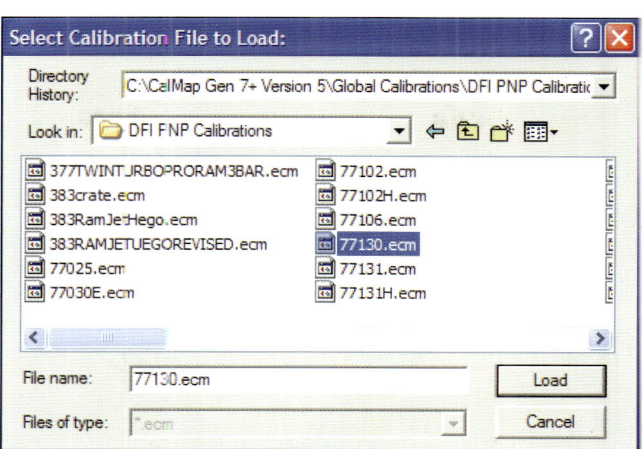

Accel DFI supplies a library of calibrations for popular crate engine combinations. This can be a real time saver when trying to get a new project up and running quickly.

legitimate temperature or weather change, and the tuner is confident in the previous day's work, the only change should be to the compensation curves. Ideally, the temperature compensation should be checked near the limits of the predicted temperature range the vehicle will see to ensure proper operation under all conditions.

Many of these aftermarket PCMs have an "Auto Tune" feature that corrects fuel delivery based on feedback from an onboard wideband oxygen sensor. For the hobbyist looking to just get something running on a car, these help reduce the tuning effort and time. For those looking to perform a proper calibration, I recommend that these be turned off. There is rarely any substitute for an accurate wideband and an experienced calibrator making the proper mathematical adjustments. I always get better results by holding the engine in steady state on a dyno and making the necessary corrections versus allowing the included strategy to "learn" the proper tune. Remember that the engine is almost never steady state while driving on the road, so the learning function never really gets a good look at any particular trim cell for very long. If the steady state calibrations are done

correctly, there will be little work for the closed loop correction later anyway. This leaves less room for error or drivability problems a week later.

Accel DFI

The Accel DFI GenVII standalone EFI system is one of the most widely used and flexible aftermarket engine controllers. Scaling of the base maps is infinitely adjustable with respect to engine speed and load. Load can be configured as in-Hg, kPa, or TPS (true alpha-N) and can be used with any standard GM MAP sensor to accommodate up to 30 psi of boost. TPS range adjustment can be conveniently done from the driver's seat by simply pressing the pedal in the calibration mode.

All base fuel calibrations are referenced from predicted air mass. Even though the actual mass flow number is not directly shown to the tuner, it is being calculated based on base VE, MAP, and modeled air temperature at the port. The result is an OEM-like calibration ability for lambda control. A properly calibrated GenVII system should drive just like a stock vehicle regardless of weather or load changes.

Accel DFI GenVII is the first widely available and affordable

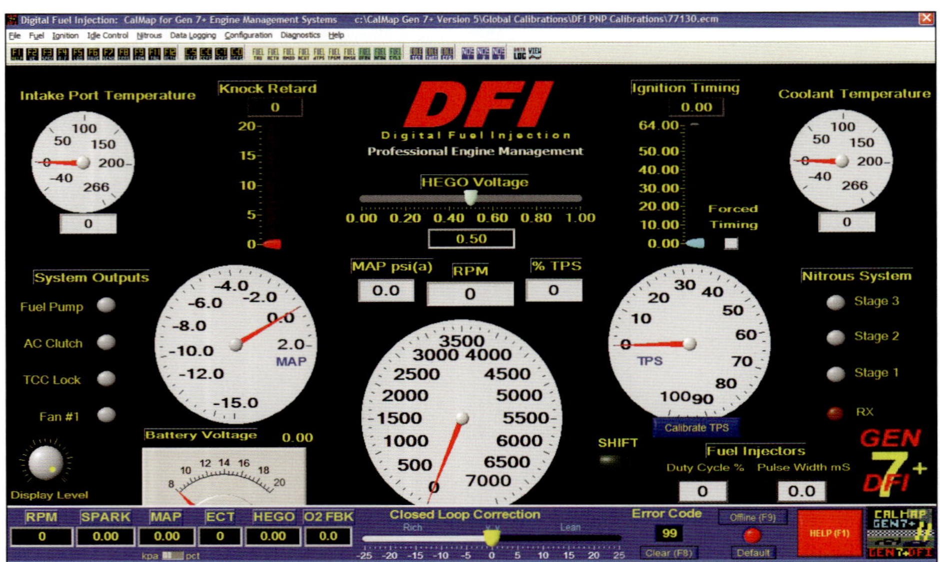

The Accel DFI system opens up initially to the virtual dashboard that shows all major sensor and engine data on one convenient screen.

recorded run with the base VE or spark maps makes it simple to see which cells were being used for the fuel and spark calculations during the run.

F.A.S.T.

Another one of the most popular aftermarket standalone EFI systems is the Fuel Air Spark Technology (F.A.S.T.) XFI system. This PCM is a

The XFI system from F.A.S.T. includes the PCM and a complete wiring harness. This makes for easy installation even where the vehicle harness is damaged or never existed in the first place.

standalone to employ true airflow modeling, including Tau (wall film) prediction. Normal acceleration enrichment tables are present for those accustomed to using them. Accel goes one step further by adding tables to represent actual wall film fueling contribution during both acceleration and deceleration with respect to speed and temperature.

The integral datalogger for the Accel GenVII is plenty capable for most WOT or drivability recording needs. Although limited to six channels, acquisition rate is fast enough to provide sufficient feedback to the calibrator. The ability to overlay a

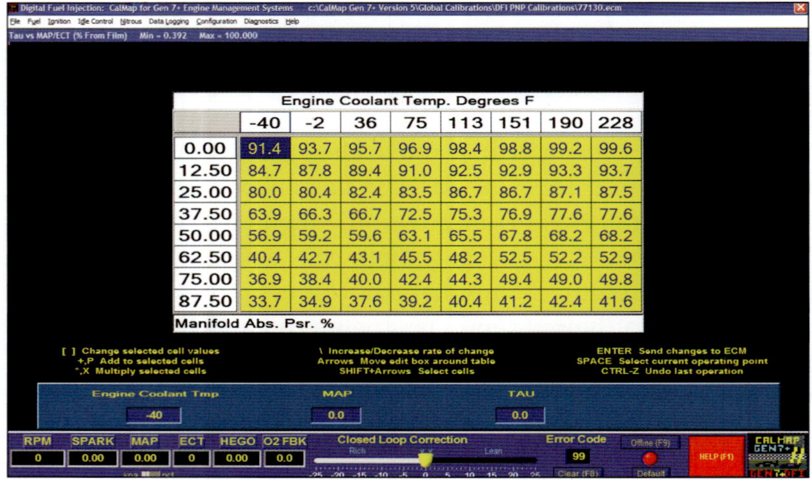

The Tau modeling in the GENVII systems is controlled mostly by the table seen here. The actual number values in the table are not so important as the differences between them. A larger difference between two points on this table represents a larger change in wall film compensation.

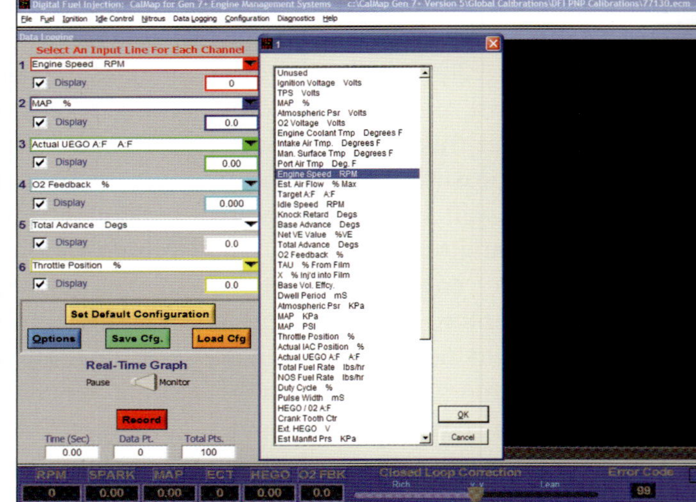

A constantly scrolling screen shows the values for up to six channels simultaneously during recording. These traces can later be overlaid upon the base fuel and spark maps to show what cells were used during a run.

With individual throttle bodies and essentially unlimited plenum volume, "stack" EFI systems like this one usually work best in alpha-N mode with a standalone PCM.

Standalone systems are replacing carburetors in more and more project vehicles. This small-block Chevy no longer needs to worry about changing jets with the weather.

true speed density system much like the Accel DFI, where an actual air mass is being calculated under all conditions. Base maps are completely user-configurable for speed and load break points at an almost infinite number of break points.

The load axis for the base VE table can be configured for MAP in inches of mercury or PSI as well as TPS for alpha-N configuration if desired. Acceleration enrichment is handled in the traditional manner with a group of tables featuring modifiers based on both MAP and TPS values for predicted fueling. The datalogger for the FAST system is equally comprehensive, with the ability to log and graph any PCM input including wideband air/fuel error.

AEM

AEM (Advanced Engine Management) offers a standalone PCM called the "Programmable Engine Management System." This system is very popular primarily with the import performance crowd. The real benefit is that AEM often packages the system together with an adaptor harness as a "plug and play" solution. This allows the user to simply replace the OEM PCM with the AEM unit and, using the pre-loaded tables for a stock vehicle, immediately start and run the car.

Because the AEM system is geared toward a wide variety of possible OEM combinations (and their vast array of sensors), it can be configured with unique individual sensor transfer functions. This means that the same PCM box can work with the engine sensors on a Toyota Supra, Honda Accord, or Mitsubishi Lancer Evolution. Calibration of the AEM system is fairly straightforward once the proper sensor configuration has been established. However,

if an incorrect transfer is selected, you are in for a long and frustrating tuning session.

The system reacts like a typical speed density PCM, and the majority of the calibration work is done in the main VE table. There are separate tables for acceleration enrichment and a whole aggregate for boost compensation to fueling. One caveat is that for serious boost levels, proper fuel tuning requires the use of their specific compensation logic. If you intend to use this system on a high-pressure application, read up on this particular part of the instruction manual closely to avoid headaches later.

Another interesting feature is the variable wastegate duty cycle control. This allows for OEM style variable control where spool up time can be reduced and boost overshoots can be controlled. The ability to tie this logic to a vehicle speed sensor effectively gives traction control to the turbocharger crowd.

MegaSquirt

A relative newcomer to the stand-alone PCM market, the MegaSquirt system is a low cost alternative to its competitors. The system is available either complete or as a do-it-yourself electronics project requiring some assembly. The MegaSquirt is a traditional speed density system much like the Accel DFI and F.A.S.T. systems.

Like the other systems, it can be configured for either alpha-N or MAP based load calculation. MegaSquirt also employs Tau-based wall film compensation for those looking for ideal tip-in fueling response. The tuning process for these PCMs is identical to other stock or aftermarket units, so experienced calibrators should not have much trouble.

Actual calibration of the unit is best handled using their MegaTune software, but can also be accomplished with a number of other PC based editors. Although some of the controls appear a bit crude compared to the more polished and established aftermarket competitors, all the right knobs are there to turn for a proper tune. For some, the difference in cost more than outweighs this minor issue.

Electromotive TEC³r

The Electromotive system is widely popular with the import performance crowd, and to an arguable extent with the domestic enthusiast as well. The system really shines with its extreme flexibility in controlling various coil pack and coil on plug ignition systems. It is a slightly more complex speed density system that packs a large amount of flexibility. However, its calibration and operation are significantly different from the traditional speed density systems previously mentioned.

The basic theory behind the operation of the Electromotive systems is their self-described "linear thermodynamics" calculation of pulsewidth. This should not be confused with a traditional thermodynamic calculation for engine performance. Rather, it is essentially their marketing terminology used to describe their pulsewidth calculation. The calculation of pulsewidth begins with a reference maximum referred to as "Gamma." The time for one Gamma (TOG) is then multiplied by a series of factors that determine the final injection time. This essentially boils down to a linearization of the injection time versus manifold pressure (MAP) with a few other factors thrown in for good measure. The net result is that the engine can be started with less initial work, but fine-tuning the system is more involved and slightly less intuitive.

It has been my experience that these systems work best when the TOG is calculated as the maximum possible mechanical injection time at redline. Since some of these systems don't allow for an injection time in excess of TOG, this ensures the full range of the injector is usable where it is needed most, at WOT near redline.

In the users' manual, VE correction is not emphasized strongly enough, as they depend heavily upon the "liner thermodynamics" to produce the proper fuel delivery. This essentially skips the step of airflow modeling, going straight from MAP sensor to injection time. As we have already seen, plenty of other engine parameters come into play in the airflow calculation, so we must still make adjustments for changes in actual VE.

Once the basic engine and sensor parameters have been properly input to the Electromotive system, I prefer to finish the calibration by adjusting the VE correction table. This table falls under the "few other factors" description mentioned above, but is critical to proper operation. The values in this table do not reflect actual engine VE, but still have a mathematical effect on fuel delivery in the same manner. Mercifully, a 5% change to this table should still yield a 5% change to delivered lambda when the injector is properly modeled for dead time, voltage offset, and flow rate. Even with this approach, an actual MAF number is never really achieved, but the overall delivered fuel ratio can remain under reasonable control.

Another very keen feature is the ability to blend TPS and MAP readings by a factor to maintain better control near idle on vehicles with large cams. Since the large camshaft can have relatively poor vacuum at and near idle, the Electromotive system allows for the partial (or complete) replacement of the MAP in the fuel equation with TPS at certain speeds. Effectively, the system switches from MAP based speed density to alpha-N as needed. This results in a more stable signal, and less variation in the fuel delivery calculation in a region that can be critical to good driving behavior.

INCA OEM Calibration Tool

The connection between the PCM and laptop is made with the ETK board, shown here installed inside the case of a prototype controller. A single cord connects to the calibrator's laptop. (Nate Tovey)

The standard of the industry today is the ETAS INCA system. INCA stands for INtegrated Calibration and Acquisition system. Used by almost every major OEM today, it allows for real-time calibration of tens of thousands of variables and tables while simultaneously recording. Cost of the system and required hardware is well beyond the budget of most aftermarket endeavors. Interface with the PCM can be configured with either an add-on module to the PCM itself or network interface over CAN bus.

Additionally, the system requires the use of what is almost always confidential data. Two files are required, the hexadecimal data record and identifier file. The data record is usually stored in the Motorola S-record format (*.s19) and identifiers as ASAM2 (*.a2l) files. The S-record contains the actual calibration and software data. The ASAM2 file is a "decoder ring" of sorts that indicates the location and size of variables in the S-record along with their conversion from hexadecimal to physical units. Only the S-record is actually stored in the PCM's memory, so acquiring the ASAM2 file is necessary to properly organize and scale the data. OEM level calibration is far more intricate than aftermarket tuning. The files used at this level usually contain tens of thousands of parameters compared to the couple hundred adjustments available in most "tuner" systems. Keeping track of this intensely detailed operation strategy and so many parameters often requires thousands of pages of reference material to accompany the actual calibration software.

Once the proper files have been loaded, INCA allows the user to view all operating parameters in table, oscilloscope, gauge, graph, alert, or numeric display. The size, range, color, and position of the display is infinitely configurable. The configuration of this data display is stored as an "experiment," and can include numerous layers to sort data sets as needed. Changes to the calibration data can be made instantaneously much like other aftermarket performance systems. INCA adds the ability to change instantly, at the touch of a button, between two completely different files while driving. This makes comparisons between two possible calibrations extremely fast and simple. Measurement recordings can be taken directly from the experiment screen and replayed later in any alternative display mode. The data acquisition rate and number of available channels is simply untouched by the traditional aftermarket tuning industry.

EXTERNAL CONTROLLERS

Modern EFI engines can often be adjusted even without actually remapping the PCM. There are a wide range of devices designed to modify PCM inputs and outputs to adjust actual performance. Some of these modify electrical signals and others mechanically change operating conditions. When used properly, these devices can be useful and cost effective tools, but it is important to understand exactly what they are doing and what their limits are.

Electronic Ignition Boxes

Ignition amplifiers are probably the most common aftermarket addition to engine electronics. These boxes amplify the primary 12-v signal to the ignition coil to upwards of 50,000 v. The result is higher spark energy and the ability to bridge either a wider spark gap or a gap through a denser charge. These are used to successfully overcome ignition blowout and misfires at WOT in most cases. Although electrical in design, their function is largely mechanical since the major impact is upon actual combustion initiation.

The MSD Digital 7 acts not only as an ignition amplifier, but also allows a degree of programmable ignition advance adjustment. (Nate Tovey)

More modern versions of these ignition boxes have added functions to shift ignition timing. The ignition box knows engine speed simply by counting input pulses. Any changes made to delivered ignition timing are unknown to the PCM. This means that the PCM may be commanding 30 degrees of advance even if the ignition box is retarding delivered timing to 25 degrees. This should be kept in mind when datalogging directly from the PCM, as actual advance needs to be calculated. A MAP sensor can be added to adjust according to manifold pressure in supercharged applications, allowing for spark retard relative to boost. This boost timing retard is usually adjusted in degrees of spark retard per pound of boost. This allows for full timing at partial load and retardation even when exceeding the maximum load calculated by the PCM. While spark advance may

The MSD BTM works in conjunction with a typical ignition amplifier. These units have a built-in two bar MAP sensor and can be adjusted for a certain number of degrees of retard for each pound of boost present. The port for the sensor can be seen in the middle here. (Nate Tovey)

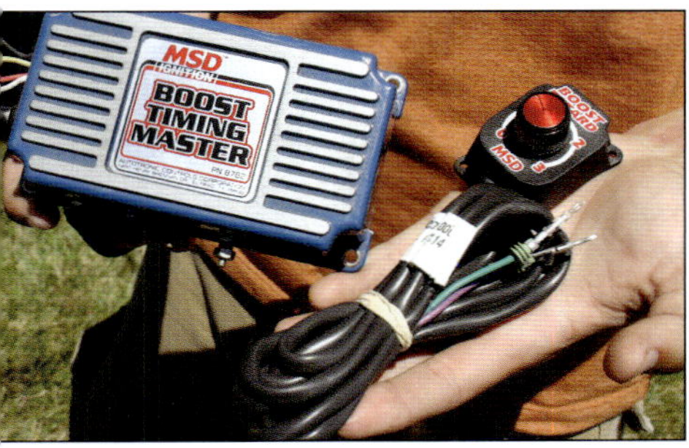

A remote knob allows user adjustability of the retard function. Up to 15 degrees of retard can be added with this controller. (Nate Tovey)

be optimized under boost, fueling needs to be increased by some method as well.

Much like the boost timing retard, other boxes allow the activation of an external trigger (such as a nitrous oxide circuit) to retard timing as well. These circuits allow for the engine to be tuned for ideal operation in naturally aspirated trim while still ensuring proper retardation during nitrous oxide injection.

VAFC / MAF Adjusters

Other devices adjust actual OEM sensor inputs to the PCM. The APEXi VAFC is one such item that intercepts and modifies the MAP sensor signal before relaying it to the PCM. By shifting actual MAP input, calculated engine load changes and the PCM makes what it considers normal adjustments. Increasing this signal causes a net increase in fuel delivery and slight drop in timing to accommodate the larger theoretical load. There are also several versions that do almost the same thing with MAF sensor signals. Since the PCM does not know that these devices are active, any adjustment to them should typically be made before mapping airflow in a full calibration to the PCM.

The MAF Translator and Diablosport MAFia converter act as voltage reduction boxes to extend the range of a given MAF sensor. Since the factory MAF reaches its 5 v maximum relatively early on GT Mustangs, the MAFia allows the sensor to feed it a higher voltage signal that is then scaled down before relaying this voltage to the PCM. This global MAF voltage reduction requires a shift in the MAF transfer function values as well as matching reductions to injector size and engine displacement to keep fueling and load calculations correct. The Diablosport ChipMaster Revolution software has a handy tool built in to speed these changes when using the MAFia. If employed correctly, a 3% change to the MAF transfer function in the editor should still result in a 3% change in delivered lambda as measured with a wideband O2 sensor on the vehicle.

More advanced MAF adjusters allow the user to completely eliminate the MAF sensor, replacing it with a generated signal. This signal is created based on inputs from engine RPM, air temperature, MAP, and stored tables for

This APEXi ECU completely replaced the factory module on this Toyota MR2. It even retains the use of a MAF sensor, making the calibration for the engine swap in this vehicle much easier.

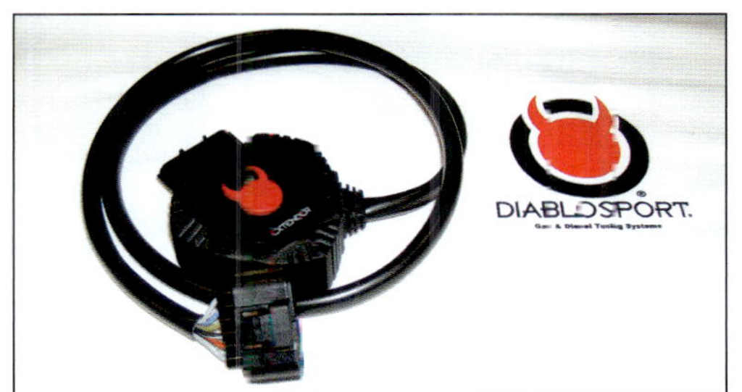

The DiabloSport MAFia acts as a scaling device on the factory MAF sensor. The meter itself is capable of outputs in excess of the PCM's 5 volt maximum. This device scales the signal down to a level that the factory PCM can still recognize. Use of this type of device requires careful recalibration of the PCM.

VE. This arrangement effectively converts the engine to speed density with the calculated air mass being delivered to the stock MAF routine in the stock PCM. This serves as a helpful method of exceeding a maximum measurement range of an undersized or restrictive OEM MAF sensor. However, one must remember that it is simply generating a simulated MAF value and all of the PCM's original strategies are still in effect. This means that scaling the simulated MAF signal down (to accommodate larger injectors) still leads to errors in calculated engine load. This can have the detrimental effect of inadvertently increasing spark advance due to a lower calculated load even when more fuel is actually being injected to match the increased airflow. Proper use of these devices on engines that are modified (especially with forced induction) almost certainly still requires changes to the PCM's calibration to perform optimally.

Piggyback Controllers

Other devices take signal modification one step further by including a

The UniChip system is a piggyback controller that intercepts and modifies the electrical signals going in and out of the factory PCM. The signals must still remain within range during closed loop to avoid setting error codes in the stock PCM.

Auxilliary injection systems include a plate that is sandwiched between the intake manifold and throttle body to fit extra injectors for WOT fueling. This example is shown on a Dodge Hemi engine as part of a supercharger kit.

full array of inputs and outputs. Systems such as the UniChip, Xede, and PMS controllers "piggyback" onto the factory PCM and are programmed to only modify signals under certain conditions. These controllers typically leave the factory PCM in control for most operating conditions, but step in to adjust fueling and ignition at high load. These piggyback programmers can be useful when attempting to operate beyond the normal engine speed limit or with "scaled" MAF or injector changes. Although fully programmable, it must be understood that the factory PCM still attempts to perform normally, whether it actually has control or not. The majority of the usefulness of these devices lies in open loop operation where the PCM is not actively trying to correct the same outputs being adjusted by the controller.

Auxiliary Injector Control

Another common electrical addition for aftermarket supercharger installations is the auxiliary injector controller (AIC). These devices are usually triggered by engine speed, manifold pressure, MAF signal, or a combination thereof to supply additional fuel, even after the factory PCM has the original injectors operating at 100% duty cycle. One or more additional injectors are positioned in the inlet tract with a variable trigger signal created by the auxiliary injector controller. Again, the factory PCM has no indication that this controller is active, so even if enough extra fuel is added from the AIC, load and fuel delivery calculations remain low. Although fueling may be corrected with the use of an AIC, ignition timing remains under PCM control and often requires some form of adjustment as well.

Placement of the additional injectors is usually less than ideal. Most such setups have the auxiliary injectors installed ahead of the throttle body, making for "wet" manifold flow. Since most EFI intake manifolds are not designed to flow both air and fuel, some distribution error is to be expected. One or more cylinders almost certainly err either rich or lean from the average indicated by the wideband sensor during testing. When using an AIC, it is recommended to inspect fuel distribution by either visual spark plug inspection or cylinder to cylinder exhaust gas temperature monitoring before declaring the final calibration "safe."

Hobbs Switches

Additional external electrical circuits can also be useful when tuning. A common addition is the Hobbs switch, a pressure activated switch used to trigger an output at a fixed boost level. These can be used to activate a second fuel pump or voltage boost under boost to provide the necessary flow volume without overheating fuel at low engine load. Hobbs switches can be used to trigger a water or alcohol injection system. For example, a supercharged engine may elect to add a 5-psi Hobbs switch for water injection to cool the air charge under high loads. If such a system is used, careful tuning must be done to find optimal spark advance both with and without the water injection. If the water tank runs dry, the engine should not be operated at full timing, so it is helpful to have a backup calibration to prevent damage. Pressure switches can also be used to deactivate a nitrous oxide spooling system on turbocharged cars.

Mechanical Adjustable Fuel Pressure Regulators

The mechanical adjustable fuel pressure regulator can be used to change the effective injector size. By placing a screw in series with the spring acting upon the flow diaphragm, pressure becomes infinitely adjustable. This can be used as a quick fix for slightly undersized fuel injectors. As shown earlier, changing from 40-psi to 50-psi fuel pressure increases effective injector size by 11.8%. Ideally, this change is done prior to any PCM calibration and the actual injector performance values are input to the tune. If the rail pressure is increased without any change to the PCM, global fuel delivery temporarily increases. Once the PCM enters closed loop operation, it learns to adjust pulsewidth down based upon feedback from the HEGOs. This is why there is seldom a long-term performance benefit from simply installing an adjustable fuel pressure regulator on modern EFI equipped vehicle.

The FMU

Many low-cost supercharger installations employ another form of fuel pressure regulator to add the necessary fueling under boost. The fuel management unit (FMU) is a rising rate regulator that increases rail pressure faster than manifold pressure. This is accomplished by exposing the diaphragm to a larger surface area of manifold pressure than fuel pressure. Installed in series with the standard fuel pressure regulator, it has no

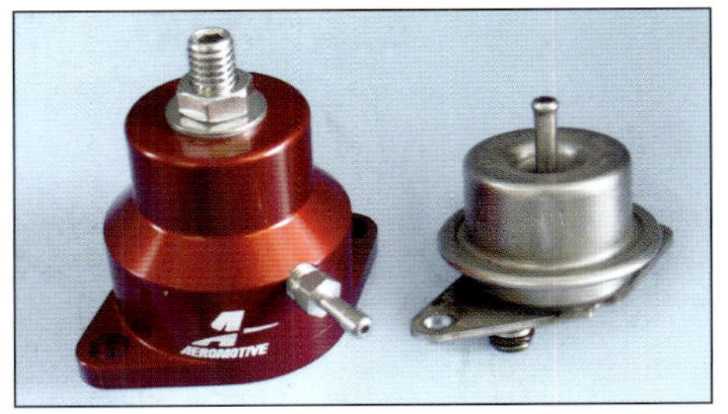

Adjustable fuel pressure regulators like the one on the left have a variable spring tension that is controlled by a setscrew. This allows base fuel pressure to be set outside of the stock range if desired. (Nate Tovey)

effect under vacuum. As manifold pressure increases, the FMU begins to progressively close off the fuel return line to raise pressure according to its rated ratio. For example, a supercharged engine making 10 psi of boost with an 8:1 FMU sees 80 psi of actual rail pressure, 70 psi of rail delta pressure, and a 32.3% increase in fuel flow above nominal if base pressure was 40 psi. This means that if the vehicle is equipped with 24 lbs/hr injectors, actual flow rate at peak boost would now be approximately 31.7 lbs/hr. While this device can add the necessary fuel enrichment for supercharged applications, spark must still be addressed separately.

Unless the PCM is equipped with a fuel rail pressure sensor (which is unlikely since these usually only exist on returnless fuel systems that don't have a return line available for FMU installation), no adjustments are being made to compensate for the added fuel delivery. This can be either good or bad depending on whether the net result is a rich or lean condition. Since the effective injector size is changing as manifold pressure changes, it makes actual airflow and fuel delivery mapping difficult at best with an FMU.

Most centrifugal supercharger and FMU installations exhibit "U" shaped delivered lambda curves where low speed and boost levels are slightly lean. As boost levels increase, effective injector size increases sharply while the PCM attempts to complete its own enrichment as commanded by its base fuel maps. At higher speeds, the stock MAF signal often reaches maximum voltage or the factory injectors reach static flow. This causes the delivered lambda to begin to lean out again. If the initial enrichment from the FMU was large enough, lambda remains

safe before reaching redline. This ever-changing lambda with respect to engine speed is less than desirable for torque and misfire. Changes to the delivered fueling from revised PCM calibration can significantly improve this. The preferred method is to install the proper injector size and recalibrate without the FMU altogether. However, if this is not possible, fuel adjustments are best made in the target lambda table or WCT enrichment function. This helps to avoid inadvertently changing ignition advance at the same time from the load shift that would otherwise occur from adjusting MAF transfer function or base VE. Be sure to verify that this change in commanded lambda does not interfere with the spark advance versus lambda table, if used.

Manual Boost Controllers and Wastegates

Cars that are equipped from the factory with a turbocharger employ some form of boost control. In the simplest installations, this is just a vacuum line from the compressor housing to the wastegate actuator. In more elegant control systems, the vacuum line is routed to an outboard boost control solenoid and then back to the wastegate actuator. Either way, enthusiasts look for ways to increase power by raising boost pressure.

One method is by simply increasing the mechanical tension on the wastegate spring. Most factory turbochargers have a threaded rod that can simply be adjusted to a shorter length to increase the preload on the same spring in the wastegate. This increased tension requires a higher boost pressure on the diaphragm to overcome it and results in increased boost and airflow to the engine as well.

Another method is to use some sort of bleed device to reduce the delivered pressure to the diaphragm of the wastegate. These devices are usually some form of "T" that is placed in the vacuum line to the wastegate actuator where one leg of the tee has a variable leak. These "fish-tank valves" are named from their common source, aquarium supply stores, where they are commonly sold. The valve is adjusted by a threaded fitting on one end. By increasing the area of the leak in the supply line, reference pressure to the wastegate is reduced relative to the manifold. The result is a wastegate that still opens at its prescribed pressure; however, that pressure is not achieved at the wastegate until a larger pressure has been realized at the manifold due to the leak between the two. Only a very small volume of air needs to escape through this valve to create the pressure difference because the volume of the vacuum lines is so small.

As with many technologies, mechanical controls have given way to electronics here as well. A number of variable boost controllers are on the market now that use an electronically controlled solenoid to open or close the boost leak on the controller. Companies like Greddy and TurboXS offer several solutions. These systems allow for the use of a feedback control circuit as well as inputs from the engine or vehicle speed to make precise adjustments to the boost level. This is often more accurate and faster than the simple bleed valves, with the added benefit of being able to record and play back boost data.

Whether the boost controller is manual or electronically controlled, care must still be taken to ensure that proper fuel delivery is achieved at the engine. Often, the PCM is completely unaware that the boost level has been increased, which has the possibility of leading to a lean condition without any correction. Additionally, some more advanced PCMs may monitor delivered MAP sensor values and trigger a fault if the delivered condition falls out of range. In some cases, it may even completely exceed the measurement range of the MAP sensor altogether, further impeding the PCM's ability to control accurate fuel delivery.

Exhaust Cutouts

Opening up the exhaust by means of a "cutout" is a simple and inexpensive way of reducing backpressure. Enthusiasts often install these as a "Y" in the exhaust system with either a removable plate or throttle style valve. Usually closed for normal operation, they are opened to allow for more total flow under race conditions. Mass air systems do not typically experience any trouble with these devices as long as they are installed far enough downstream from the HEGO to avoid detecting the fresh air wash. The increased system airflow is simply measured by the MAF sensor, and the PCM reacts normally. Speed density systems will likely experience a more noticeable change in performance. The reduced backpressure raises engine VE when opened. Without any recalibration, the PCM assumes normal operation and delivers what it considers the proper amount of fuel based on MAP and speed. The result can be a lean condition that may lead to knock.

I once had a speed density engine that was calibrated on an engine

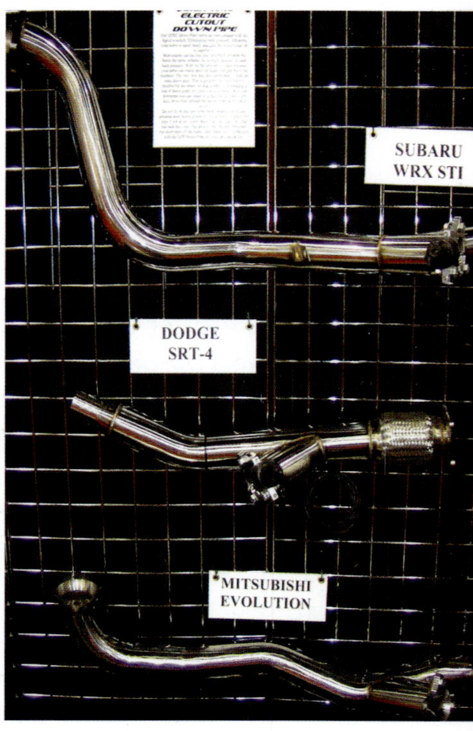

Exhaust cutouts have an instant effect upon flow and volumetric efficiency. This particular unit is electronically controlled by a switch inside the vehicle.

dynamometer using the normal long tube headers open to the dyno exhaust. In this low backpressure configuration, the engine made about 380 hp. This same engine was later installed in the vehicle with a closed exhaust, catalysts, and mufflers. The added backpressure had reduced total flow to the point where the engine ran rich and would not tolerate the same ignition lead. The VE tables required remapping to take as much as 15% out of the WOT values, and timing needed to be retarded because of the added backpressure. The result was a more disappointing 320 hp, but with a safe, emissions friendly tune. To install a cutout on this vehicle would almost certainly allow the engine to breathe again at the 380-hp level, but the PCM would not know to add the appropriate fuel to compensate. The moral of the story is: If a cutout is used with speed density, be prepared to recalibrate or risk engine performance and durability.